SpringerBriefs in Mathematical Physics

Volume 11

More information about this series at http://www.springer.com/series/11953

Makoto Katori

Bessel Processes, Schramm–Loewner Evolution, and the Dyson Model

Springer

Makoto Katori
Department of Physics
Chuo University
Tokyo
Japan

Additional material to this book can be downloaded from http://extras.springer.com.

ISSN 2197-1757 ISSN 2197-1765 (electronic)
SpringerBriefs in Mathematical Physics
ISBN 978-981-10-0274-8 ISBN 978-981-10-0275-5 (eBook)
DOI 10.1007/978-981-10-0275-5

Library of Congress Control Number: 2015959919

Printed on acid-free paper

This Springer imprint is published by SpringerNature
The registered company is Springer Science+Business Media Singapore Pte Ltd.

To Hiroko, Machiko, and Rieko

Preface

This book is based on my graduate-course lectures given at the Graduate School of Mathematics of the University of Tokyo in October 2008 (at the invitation of T. Funaki and M. Jimbo), at the Department of Physics of the University of Tokyo in November 2010 (at the invitation of S. Miyashita), at the Department of Mathematics of Tokyo Institute of Technology in December 2010 (at the invitation of K. Uchiyama), at École de Physique des Houches (Les Houches Physics School) in May 2011 (organized by C. Donati-Martin, S. Péché and G. Schehr), at the Faculty of Mathematics of Kyushu University in June 2013 (at the invitation of H. Osada and T. Shirai), and at the Graduate School of Arts and Sciences of the University of Tokyo in July 2014 (at the invitation of A. Shimizu). First of all I would like to thank those organizers for giving me such opportunities.

The purpose of my lectures is to introduce recent topics in mathematical physics and probability theory, especially the topics on the Schramm–Loewner evolution (SLE) and interacting particle systems related to random matrix theory. A typical example of the latter systems is Dyson's Brownian motion model. For this purpose I have considered one story to tell the SLE and the Dyson model as 'children' of the Bessel processes. The Bessel processes make a one-parameter family of one-dimensional diffusion processes with parameter D, in which the D-dimensional Bessel process, $\mathrm{BES}^{(D)}$, is defined as the radial part of the D-dimensional Brownian motion, if D is a positive integer. This definition implies that Bessel processes are 'children' of the Brownian motion, and hence, the SLE and the Dyson model are 'grandchildren' of the Brownian motion.

The organization of this book is very simple. In Chap. 1 the parenthood of Brownian motion in diffusion processes is clarified and we define $\mathrm{BES}^{(D)}$ for any $D \geq 1$. There, the importance of two aspects of $\mathrm{BES}^{(3)}$ is explained. SLE is introduced as a complexification of $\mathrm{BES}^{(D)}$ in Chap. 2. We show that rich mathematics and physics involved in SLE are due to the nontrivial dependence of the Bessel flow on D. In Chap. 3 Dyson's Brownian motion model with parameter β is introduced as a multivariate extension of $\mathrm{BES}^{(D)}$ with the relation $D = \beta + 1$. We will concentrate on the case where $\beta = 2$. In this case the Dyson model inherits the two aspects of $\mathrm{BES}^{(3)}$ and has very strong solvability. That is, the process is proved to be determinantal in the sense that all spatio-temporal correlation functions are given by determinants, and all of them are controlled by a single function called the correlation kernel.

Many parts of this book come from the joint work with Hideki Tanemura. I thank him very much for the fruitful collaboration over 10 years. I would like to thank Alexei Borodin, John Cardy, Patrik Ferrari, Peter John Forrester, Piotr Graczyk, Kurt Johansson, Takashi Imamura, Christian Krattenthaler, Takashi Kumagai, Neil O'Connell, Hirofumi Osada, Tomohiro Sasamoto, Grégory Schehr, Tomoyuki Shirai, and Craig Tracy for giving me encouragement to prepare the manuscript. I am grateful to Nizar Demni, Sergio Andraus, Syota Esaki, Ryoki Fukushima, and Shuta Nakajima for careful reading of the draft and a lot of useful comments. All suggestions given by two anonymous reviewers of this book are very important and useful for improving the text and I acknowledge their efforts very much. Thanks are due to Naoki Kobayashi and Kan Takahashi for preparing several figures in the book.

I thank Masayuki Nakamura at the Editorial Department of Springer Japan for his truly kind assistance during the preparation of this manuscript.

The research of the author was supported in part by the Grant-in-Aid for Scientific Research (C) (No.21540397 and No.26400405) of the Japan Society for the Promotion of Science.

Tokyo
December 2015

Makoto Katori

Contents

Chapter 1
Bessel Processes

Abstract Basic notions of probability theory and stochastic analysis are explained using the Brownian motion and its functionals. The concepts of probability space and filtration are introduced, and the strong Markov property and the martingale property of Brownian motion are explained. A diffusion process is defined as a stochastic process having a continuous path almost surely with the strong Markov property. Quadratic variations and stochastic integrals are defined and Itô's formula is given which enables us to derive stochastic differential equations (SDEs) for diffusion processes. A D-dimensional Bessel process is originally defined as the distance from the origin of a D-dimensional Brownian motion for positive integers of D. By giving the SDE and the transition probability density, we show that it can be defined for all real values $D \geq 1$. Concerning the behavior of the Bessel flow, there exist two critical dimensions, $D_c = 2$ and $\overline{D}_c = 3/2$. For the three-dimensional Bessel process, its two aspects are emphasized: Aspect 1 as a radial part of the three-dimensional Brownian motion, and Aspect 2 as a one-dimensional Brownian motion conditioned to stay positive.

1.1 One-Dimensional Brownian Motion (BM)

We consider the motion of a Brownian particle in one-dimensional space $\mathbb{R}$ starting from the origin 0 at time $t = 0$. At each time $t > 0$, the particle position is randomly distributed, and each realization of its path (trajectory) is denoted by ω and called a *sample path* or simply a *path*. Let Ω be the collection of all sample paths and we call it the *sample path space*. The position of the Brownian particle at time $t \geq 0$ in a path $\omega \in \Omega$ is written as $B(t, \omega)$. Usually we omit ω and simply write it as $B(t),\ t \geq 0$.

We represent each event associated with the process by a subset of Ω, and the collection of all events is denoted by $\mathscr{F}$. The whole sample path space Ω and the empty set $\emptyset$ are in $\mathscr{F}$. For any two sets $\mathsf{A}, \mathsf{B} \in \mathscr{F}$, we assume that $\mathsf{A} \cup \mathsf{B} \in \mathscr{F}$ and $\mathsf{A} \cap \mathsf{B} \in \mathscr{F}$. If $\mathsf{A} \in \mathscr{F}$, then its complement A^c is also in $\mathscr{F}$. It is closed for any infinite sum of events in the sense that, if $\mathsf{A}_n \in \mathscr{F}, n = 1, 2, \dots$, then $\cup_{n \geq 1} \mathsf{A}_n \in \mathscr{F}$. Such a collection is said to be a *σ-field* (sigma-field). The symbol σ is for 'sum'.

M. Katori, *Bessel Processes, Schramm–Loewner Evolution, and the Dyson Model*, SpringerBriefs in Mathematical Physics 11, DOI 10.1007/978-981-10-0275-5_1

A *probability measure* P is a nonnegative function defined on the σ-field $\mathscr{F}$. Since any element of $\mathscr{F}$ is given by a set as above, any input of P is a set; P is a *set function*. It satisfies the following properties: $\mathrm{P}[\mathsf{A}] \geq 0$ for all $\mathsf{A} \in \mathscr{F}$, $\mathrm{P}[\Omega] = 1$, $\mathrm{P}[\emptyset] = 0$, and if $\mathsf{A}, \mathsf{B} \in \mathscr{F}$ are disjoint, $\mathsf{A} \cap \mathsf{B} = \emptyset$, then $\mathrm{P}[\mathsf{A} \cup \mathsf{B}] = \mathrm{P}[\mathsf{A}] + \mathrm{P}[\mathsf{B}]$. In particular, $\mathrm{P}[\mathsf{A}^c] = 1 - \mathrm{P}[\mathsf{A}]$ for all $\mathsf{A} \in \mathscr{F}$. The triplet $(\Omega, \mathscr{F}, \mathrm{P})$ is called the *probability space*.

The smallest σ-field containing all intervals on $\mathbb{R}$ is called the Borel σ-field and denoted by $\mathscr{B}(\mathbb{R})$. A *random variable* or *measurable function* is a real-valued function $f(\omega)$ on Ω such that, for every Borel set $A \in \mathscr{B}(\mathbb{R})$, $f^{-1}(A) \in \mathscr{F}$. Two events A and B are said to be *independent* if $\mathrm{P}[\mathsf{A} \cap \mathsf{B}] = \mathrm{P}[\mathsf{A}]\mathrm{P}[\mathsf{B}]$. Two random variables X and Y are *independent* if the events $\mathsf{A} = \{X : X \in A\}$ and $\mathsf{B} = \{Y : Y \in B\}$ are independent for any $A, B \in \mathscr{B}(\mathbb{R})$.

The *one-dimensional standard Brownian motion*, $\{B(t, \omega) : t \geq 0\}$, has the following three properties:

(BM1) $B(0, \omega) = 0$ with probability one.

(BM2) There is a subset of the sample path space $\widetilde{\Omega} \subset \Omega$, such that $\mathrm{P}[\widetilde{\Omega}] = 1$ and for any $\omega \in \widetilde{\Omega}$, $B(t, \omega)$ is a real continuous function of t. We say that $B(t)$ has a *continuous path* almost surely (a.s., for short).

(BM3) For an arbitrary $M \in \mathbb{N} \equiv \{1, 2, 3, \dots\}$, and for any sequence of times, $t_0 \equiv 0 < t_1 < \cdots < t_M$, the increments $B(t_m) - B(t_{m-1})$, $m = 1, 2, \dots, M$, are independent, and each increment is in *the normal distribution* (*the Gaussian distribution*) with mean 0 and variance $\sigma^2 = t_m - t_{m-1}$. It means that for any $1 \leq m \leq M$ and $a < b$,

$$\mathrm{P}[B(t_m) - B(t_{m-1}) \in [a, b]] = \int_a^b p(t_m - t_{m-1}, z|0)dz,$$

where we define for $x, y \in \mathbb{R}$

$$p(t, y|x) = \begin{cases} \dfrac{1}{\sqrt{2\pi t}} e^{-(x-y)^2/2t}, & \text{for } t > 0, \\ \delta(x - y), & \text{for } t = 0. \end{cases} \tag{1.1}$$

Unless otherwise noted, the one-dimensional standard Brownian motion is simply abbreviated to BM in this lecture note. The probability measure P for the BM in particular is called the *Wiener measure*. The expectation with respect to the probability measure P is denoted by E. We write the conditional probability as $\mathrm{P}[\cdot|C]$, where C denotes the condition. The conditional expectation is similarly written as $\mathrm{E}[\cdot|C]$.

The third property **(BM3)** given above implies that for any $0 \leq s \leq t < \infty$

$$\mathrm{P}[B(t) \in A | B(s) = x] = \int_A p(t - s, y|x)dy \tag{1.2}$$

holds, $^\forall A \in \mathscr{B}(\mathbb{R}), ^\forall x \in \mathbb{R}$. Therefore the integral kernel $p(t, y|x)$ given by (1.1) is called the *transition probability density function* of Brownian motion starting from x. The probability that the BM is observed in a region $A_m \in \mathscr{B}(\mathbb{R})$ at time t_m for each $m = 1, 2, \ldots, M$ is then given by

$$P[B(t_m) \in A_m, m = 1, 2, \ldots, M] = \int_{A_1} dx_1 \cdots \int_{A_M} dx_M \prod_{m=1}^{M} p(t_m - t_{m-1}, x_m|x_{m-1}), \tag{1.3}$$

where $x_0 \equiv 0$.

By **(BM3)**, we can see that, for any $c > 0$, the probability distribution of $B(c^2t)/c$ is equivalent to that of $B(t)$ at arbitrary time $t \geq 0$. It is written as

$$\frac{1}{c}B(c^2t) \stackrel{\mathrm{d}}{=} B(t), \quad ^\forall c > 0,$$

where the symbol $\stackrel{\mathrm{d}}{=}$ is for *equivalence in distribution*. Moreover, (1.3) implies that, for any $c > 0$, $B(t)$, $t \geq 0$ and its time-changed process with $t \mapsto c^2t$ multiplied by a factor $1/c$ (dilatation) follow the same probability law. This *equivalence in probability law* of stochastic processes is expressed as

$$(B(t))_{t\geq 0} \stackrel{\text{(law)}}{=} \left(\frac{1}{c}B(c^2t)\right)_{t\geq 0}, \quad ^\forall c \geq 0, \tag{1.4}$$

and called the *scaling property of Brownian motion*.

For $a > 0$, let $T_a = \inf\{t \geq 0 : B(t) = a\}$. Then for any $t \geq 0$,

$$P[T_a < t, B(t) < a] = P[T_a < t, B(t) > a], \tag{1.5}$$

since the transition probability density (1.1) is a symmetric function of the increment $y - x$. This property is called the *reflection principle* of BM. For $\{\omega : B(t) > a\} \subset \{\omega : T_a < t\}$, $a > 0$, the above is equal to $P[B(t) > a]$.

The formula (1.3) also means that for any fixed $s \geq 0$, under the condition that $B(s)$ is given, $\{B(t) : t \leq s\}$ and $\{B(t) : t > s\}$ are independent. This independence of the events in the future and those in the past is called the *Markov property*. A positive random variable τ is called *stopping time* (or *Markov time*), if the event $\{\omega : \tau \leq t\}$ is determined by the behavior of the process until time t and independent of that after t. For any stopping time τ, $\{B(t) : t \leq \tau\}$ and $\{B(t) : t > \tau\}$ are independent. It is called the *strong Markov property*. A stochastic process which has the strong Markov property and has a continuous path almost surely is generally called a *diffusion process*.

For each time $t \in [0, \infty)$, we write the smallest σ-field generated by the BM up to time $t \geq 0$ as $\sigma(B(s) : 0 \leq s \leq t)$ and define

$$\mathscr{F}_t \equiv \sigma(B(s) : 0 \leq s \leq t), \quad t \geq 0. \tag{1.6}$$

By definition, with respect to any event in $\mathscr{F}_t$, $B(s)$ is measurable at every $s \in [0, t]$. Then we have a nondecreasing family $\{\mathscr{F}_t : t \geq 0\}$ of sub-σ-fields of the original σ-field $\mathscr{F}$ in the probability space $(\Omega, \mathscr{F}, \mathrm{P})$ such that $\mathscr{F}_s \subset \mathscr{F}_t \subset \mathscr{F}$ for $0 \leq s < t < \infty$. We call this family of σ-fields a *filtration*.

The BM started at $x \in \mathbb{R}$, which is denoted by $B^x(t),\ t \geq 0$, is defined by

$$B^x(t) = x + B(t), \quad x \in \mathbb{R}, \quad t \geq 0. \tag{1.7}$$

We define $\mathrm{P}^x[B(t) \in \cdot] = \mathrm{P}[B^x(t) \in \cdot]$ and $\mathrm{E}^x[f(B(t))] = \mathrm{E}[f(B^x(t))]$ for any bounded measurable function f, $t \geq 0$. The stopping time τ mentioned above can be defined using the notion of filtration as follows: $\{\omega : \tau \leq t\} \in \mathscr{F}_t, {}^\forall t \geq 0$. The strong Markov property of BM is now expressed as

$$\mathrm{E}[f(B(s+t))|\mathscr{F}_s] = \mathrm{E}^{B(s)}[f(B(t))], \quad t \geq 0, \quad \text{a.s.}, \tag{1.8}$$

provided that $s \geq 0$ is any realization of a stopping time τ and f is an arbitrary measurable bounded function.

Since the probability density of increment in any time interval $t - s > 0$, $p(t - s, z|0)$, has mean zero, BM satisfies the equality

$$\mathrm{E}[B(t)|\mathscr{F}_s] = B(s), \quad 0 \leq s < t < \infty, \quad \text{a.s.} \tag{1.9}$$

That is, the mean is constant in time, even though the variance increases in time as $\sigma^2 = t$. Processes with such a property are called *martingales*. We note that for $0 \leq s < t < \infty$,

$$\begin{aligned}
\mathrm{E}[B(t)^2|\mathscr{F}_s] &= \mathrm{E}[(B(t) - B(s))^2 + 2(B(t) - B(s))B(s) + B(s)^2|\mathscr{F}_s] \\
&= \mathrm{E}[(B(t) - B(s))^2|\mathscr{F}_s] + 2\mathrm{E}[(B(t) - B(s))B(s)|\mathscr{F}_s] + \mathrm{E}[B(s)^2|\mathscr{F}_s].
\end{aligned}$$

By the property **(BM3)** and the definition of $\mathscr{F}_s$,

$$\begin{aligned}
&\mathrm{E}[(B(t) - B(s))^2|\mathscr{F}_s] = t - s, \\
&\mathrm{E}[(B(t) - B(s))B(s)|\mathscr{F}_s] = \mathrm{E}[B(t) - B(s)|\mathscr{F}_s]\, B(s) = 0, \\
&\mathrm{E}[B(s)^2|\mathscr{F}_s] = B(s)^2.
\end{aligned}$$

Then we have the equality

$$\mathrm{E}[B(t)^2 - t|\mathscr{F}_s] = B(s)^2 - s, \quad 0 \leq s < t < \infty, \quad \text{a.s.} \tag{1.10}$$

It means that $B(t)^2 - t$ is a martingale. (See Exercise 1.1.)

For the transition probability density of BM (1.1), it should be noted that $p(\cdot, y|x) = p(\cdot, x|y)$ for any $x, y \in \mathbb{R}$, and $u_t(x) \equiv p(t, y|x)$ is a unique solution of the *heat equation* (*diffusion equation*)

$$\frac{\partial}{\partial t}u_t(x) = \frac{1}{2}\frac{\partial^2}{\partial x^2}u_t(x), \quad x \in \mathbb{R}, \quad t \geq 0 \tag{1.11}$$

with the initial condition $u_0(x) = \delta(x-y)$. The solution of (1.11) with the initial condition $u_0^f(x) = f(x),\ x \in \mathbb{R}$ is then given by

$$u_t^f(x) = \mathrm{E}^x[f(B(t))] = \int_{-\infty}^{\infty} f(y)p(t, y|x)dy, \tag{1.12}$$

if f is a measurable function satisfying the condition $\int_{-\infty}^{\infty} e^{-ax^2}|f(x)|dx < \infty$ for some $a > 0$. Since $p(t, y|x)$ plays as an integral kernel in (1.12), it is also called the *heat kernel*.

For $0 \leq s < t < \infty$, $\xi \in \mathbb{R}$, consider $\mathrm{E}[e^{\sqrt{-1}\xi(B(t)-B(s))}|\mathscr{F}_s]$. Using p, it is calculated as follows:

$$\int_{-\infty}^{\infty} e^{\sqrt{-1}\xi z}p(t-s, z|0)dz = \int_{-\infty}^{\infty} e^{\sqrt{-1}\xi z}\frac{e^{-z^2/2(t-s)}}{\sqrt{2\pi(t-s)}}dz$$
$$= e^{-\xi^2(t-s)/2}.$$

The obtained function of $\xi \in \mathbb{R}$,

$$\mathrm{E}[e^{\sqrt{-1}\xi(B(t)-B(s))}|\mathscr{F}_s] = e^{-\xi^2(t-s)/2}, \quad 0 \leq s < t < \infty, \tag{1.13}$$

is called the *characteristic function* of BM.

1.2 Martingale Polynomials of BM

For BM, we perform the following transformation with the parameter $\alpha \in \mathbb{C} \equiv \{z = x + \sqrt{-1}y : x, y \in \mathbb{R}\}$, $B \mapsto \check{B}_\alpha$,

$$\check{B}_\alpha(t) = \frac{e^{\alpha B(t)}}{\mathrm{E}[e^{\alpha B(t)}]}, \quad t \geq 0, \tag{1.14}$$

which is called the *Esscher transformation*. It is easy to see that

$$\mathrm{E}[e^{\alpha B(t)}] = \int_{-\infty}^{\infty} e^{\alpha x}p(t, x|0)dx = e^{\alpha^2 t/2}, \quad t \geq 0.$$

Then the above is written as

$$\check{B}_\alpha(t) = G_\alpha(t, B(t)), \quad t \geq 0$$

with

$$G_\alpha(t,x) = e^{\alpha x - \alpha^2 t/2}. \tag{1.15}$$

For $0 < s < t$,

$$\begin{aligned}
\mathrm{E}[G_\alpha(t, B(t))|\mathscr{F}_s] &= \frac{\mathrm{E}[e^{\alpha B(t)}|\mathscr{F}_s]}{\mathrm{E}[e^{\alpha B(t)}]} \\
&= \frac{\mathrm{E}[e^{\alpha B(s)} e^{\alpha(B(t)-B(s))}|\mathscr{F}_s]}{\mathrm{E}[e^{\alpha B(s)} e^{\alpha(B(t)-B(s))}]}.
\end{aligned}$$

By the definition of $\mathscr{F}_s$ and independence of increment of BM (the property **(BM3)**), the numerator is equal to $e^{\alpha B(s)}\mathrm{E}[e^{\alpha(B(t)-B(s))}]$, and the denominator is equal to $\mathrm{E}[e^{\alpha B(s)}]\mathrm{E}[e^{\alpha(B(t)-B(s))}]$. Hence the above equals $e^{\alpha B(s)}/\mathrm{E}[e^{\alpha B(s)}] = G_\alpha(s, B(s))$. Therefore, $G_\alpha(t, B(t))$ is a martingale:

$$\mathrm{E}[G_\alpha(t, B(t))|\mathscr{F}_s] = G_\alpha(s, B(s)), \quad 0 \le s \le t. \tag{1.16}$$

The function (1.15) is expanded as

$$G_\alpha(t,x) = \sum_{n=0}^{\infty} m_n(t,x) \frac{\alpha^n}{n!} \tag{1.17}$$

with

$$m_n(t,x) = \left(\frac{t}{2}\right)^{n/2} H_n\left(\frac{x}{\sqrt{2t}}\right), \quad n \in \mathbb{N}_0 \equiv \{0, 1, 2, \dots\}. \tag{1.18}$$

Here $\{H_n(x)\}_{n\in\mathbb{N}_0}$ are the *Hermite polynomials* of degrees $n \in \mathbb{N}_0$,

$$H_n(x) = (-1)^n e^{x^2} \frac{d^n e^{-x^2}}{dx^n} \tag{1.19}$$

$$= \sum_{k=0}^{[n/2]} (-1)^k \frac{n!}{k!(n-2k)!} (2x)^{n-2k}, \tag{1.20}$$

where for $a \ge 0$, $[a]$ denotes the largest integer which is not larger than a (see Exercises 1.2–1.4).

Lemma 1.1 *The functions* $\{m_n(t,x)\}_{n\in\mathbb{N}_0}$ *satisfy the following.*
(i) They are monic polynomials of degrees $n \in \mathbb{N}_0$ *with time-dependent coefficients:*

$$m_n(t,x) = x^n + \sum_{k=0}^{n-1} c_n^{(k)}(t) x^k, \quad t \ge 0.$$

(ii) For $0 \le k \le n-1$, $c_n^{(k)}(0) = 0$. *That is,*

$$m_n(0, x) = x^n, \quad n \in \mathbb{N}_0.$$

(iii) If we set $x = B(t)$, *they provide martingales:*

$$\mathrm{E}[m_n(t, B(t))|\mathscr{F}_s] = m_n(s, B(s)), \quad 0 \le s \le t, \quad n \in \mathbb{N}_0. \tag{1.21}$$

Proof By the definition (1.18) with (1.20), (i) and (ii) are obvious. Note that when n is even (resp. odd), $c_n^{(k)}(t) \equiv 0$ for odd (resp. even) k. Since $G_\alpha(t, B(t)),\ t \ge 0$ was shown to be a martingale for any $\alpha \in \mathbb{C}$, the expansion (1.17) implies (iii). □

We call $\{m_n(t, x)\}_{n \in \mathbb{N}_0}$ the *fundamental martingale polynomials* associated with BM [4] (see also [8]). For $n = 2$, (1.20) gives $H_2(x) = 4x^2 - 2$, and then $m_2(t, x) = x^2 - t$ by (1.18). We already proved in (1.10) that $m_2(t, B(t)) = B(t)^2 - t$ is a martingale.

The Fourier transformation of $G_\alpha(t, x)$ with respect to the parameter $\alpha \in \mathbb{R}$ is calculated as

$$\widehat{G}_w(t, x) \equiv \int_{-\infty}^{\infty} \frac{e^{-\sqrt{-1}\alpha w}}{2\pi} G_\alpha(t, x) d\alpha = \frac{e^{-(\sqrt{-1}x + w)^2/2t}}{\sqrt{2\pi t}}. \tag{1.22}$$

Owing to the factor $e^{-w^2/2t}$ in $\widehat{G}_w(t, x)$, the following calculations are justified,

$$\begin{aligned} G_\alpha(t, x) &= \int_{-\infty}^{\infty} e^{\sqrt{-1}\alpha w} \widehat{G}_w(t, x) dw \\ &= \sum_{n=0}^{\infty} \frac{\alpha^n}{n!} \int_{-\infty}^{\infty} (\sqrt{-1}w)^n \widehat{G}_w(t, x) dw, \end{aligned} \tag{1.23}$$

which proves the integral representation of $m_n(t, x)$,

$$\begin{aligned} m_n(t, x) &= \int_{-\infty}^{\infty} (\sqrt{-1}w)^n \widehat{G}_w(t, x) dw \\ &= \int_{-\infty}^{\infty} (\sqrt{-1}w)^n \frac{e^{-(\sqrt{-1}x + w)^2/2t}}{\sqrt{2\pi t}} dw, \quad t \ge 0, \quad n \in \mathbb{N}_0. \end{aligned} \tag{1.24}$$

We define this type of integral transformation of a function f as

$$\mathscr{I}[f(W)|(t, x)] = \int_{-\infty}^{\infty} f(\sqrt{-1}w) \widehat{G}_w(t, x) dw. \tag{1.25}$$

Then the above results are written as

$$m_n(t, x) = \mathscr{I}[W^n|(t, x)], \quad t \ge 0, \quad n \in \mathbb{N}_0. \tag{1.26}$$

In (1.24), we change the integral variable $w \to y$ by $y = \sqrt{-1}x + w$. Since the integrand is an entire function of y, the obtained integral on $y \in (-\infty+ix, \infty+ix)$ is equal to that on $y \in \mathbb{R}$ by Cauchy's integral theorem. For $\widehat{G}_{y-\sqrt{-1}x}(t,x) = p(t,y|0)$ with (1.1), we have

$$m_n(t,x) = \int_{-\infty}^{\infty} (x+\sqrt{-1}y)^n p(t,y|0)dy, \quad t \geq 0, \quad n \in \mathbb{N}_0.$$

Here we introduce BM, $\widetilde{B}(t),\ t \geq 0$, which is independent of $B(t),\ t \geq 0$ and whose probability space is written as $(\widetilde{\Omega}, \widetilde{\mathscr{F}}, \widetilde{\mathrm{P}})$ with expectation $\widetilde{\mathrm{E}}$ and the filtration $\widetilde{\mathscr{F}}_t = \sigma(\widetilde{B}(s) : 0 \leq s \leq t),\ t \geq 0$. Then the above is written as

$$m_n(t,x) = \widetilde{\mathrm{E}}[(x+\sqrt{-1}\widetilde{B}(t))^n], \quad t \geq 0, \quad n \in \mathbb{N}_0. \tag{1.27}$$

1.3 Drift Transformation

Let b be a real constant and consider the *drifted Brownian motion*

$$B^{(b)}(t) = B(t) + bt, \quad t \geq 0. \tag{1.28}$$

The constant b is called a *drift coefficient*. Its transition probability density should be obtained by performing the *Galilean transformation* of (1.1) as

$$\begin{aligned} p(t, y-bt|x) &= \frac{1}{\sqrt{2\pi t}} e^{-\{x-(y-bt)\}^2/2t} \\ &= e^{b(y-x)-b^2t/2} p(t,y|x). \end{aligned}$$

We regard this as a transformation of function

$$p(t,y|x) \mapsto p^{(b)}(t,y|x) \equiv e^{b(y-x)-b^2t/2} p(t,y|x) \tag{1.29}$$

and call it the *drift transformation*.

Let $\mathrm{P}^{(b)}$ be the probability law with respect to the drifted Brownian motion (1.28) and $\mathrm{E}^{(b)}$ be the expectation. Consider the characteristic function of the process

$$X(t) \equiv B^{(b)}(t) - bt, \quad t \geq 0. \tag{1.30}$$

For $\xi \in \mathbb{R}, 0 \leq s < t < \infty$, it is given by

$$\begin{aligned}
&\mathrm{E}^{(b)}[e^{\sqrt{-1}\xi(X(t)-X(s))}|\mathscr{F}_s] \\
&= \int_{-\infty}^{\infty} e^{\sqrt{-1}\xi\{z-b(t-s)\}} p^{(b)}(t-s, z|0)dz \\
&= \int_{-\infty}^{\infty} e^{\sqrt{-1}\xi\{z-b(t-s)\}} e^{bz-b^2(t-s)/2} p(t-s, z|0)dz. \qquad (1.31)
\end{aligned}$$

If we insert (1.1) and perform the Gaussian integral, we have

$$\mathrm{E}^{(b)}[e^{\sqrt{-1}\xi(X(t)-X(s))}|\mathscr{F}_s] = e^{-\xi^2(t-s)/2}. \qquad (1.32)$$

It is the same as (1.13) and hence it implies that $X(t),\ t \geq 0$ is a Brownian motion under $\mathrm{P}^{(b)}$. It should be so, since (1.28) and (1.30) give $X(t) = (B(t) + bt) - bt = B(t),\ t \geq 0$. On the other hand, the last line of (1.31) is rewritten as

$$\mathrm{E}\left[e^{\sqrt{-1}\xi\{(B(t)-bt)-(B(s)-bs)\}} \frac{e^{bB(t)-b^2t/2}}{e^{bB(s)-b^2s/2}} \middle| \mathscr{F}_s \right], \qquad (1.33)$$

since p is the transition probability density of BM, $B(t),\ t \geq 0$. The equivalence between (1.32) and (1.33) implies the following statement: for any $\mathscr{F}_t$-measurable bounded function $F, 0 \leq s \leq t < \infty$,

$$\mathrm{E}^{(b)}[F(B^{(b)}(t))|\mathscr{F}_s] = \mathrm{E}\left[F(B(t)) \frac{e^{bB(t)-b^2t/2}}{e^{bB(s)-b^2s/2}} \middle| \mathscr{F}_s \right]. \qquad (1.34)$$

By (1.16), $G_{\alpha+b}(t, B(t)),\ t \geq 0$ is martingale for an arbitrary $\alpha \in \mathbb{C}$. We see that

$$\begin{aligned}
G_{\alpha+b}(t,x) &= e^{(\alpha+b)x-(\alpha+b)^2t/2} \\
&= e^{bx-b^2t/2} G_\alpha(t, x-bt) \\
&= \sum_{m=0}^{\infty} e^{bx-b^2t/2} m_n(t, x-bt) \frac{\alpha^n}{n!}.
\end{aligned}$$

On the other hand, by the definition (1.25), we see that

$$\begin{aligned}
\sum_{n=0}^{\infty} \mathscr{I}[e^{bW} W^n|(t,x)] \frac{\alpha^n}{n!} &= \mathscr{I}[e^{(\alpha+b)W}|(t,x)] \\
&= \int_{-\infty}^{\infty} e^{\sqrt{-1}(\alpha+b)w} \widehat{G}_w(t,x)dw \\
&= G_{\alpha+b}(t,x),
\end{aligned}$$

where (1.23) is used. Then, if we set

$$\begin{aligned} m_n^{(b)}(t,x) &\equiv e^{bx-b^2t/2} m_n(t, x-bt) \\ &= \mathscr{I}[e^{bW} W^n | (t,x)], \quad t \geq 0, \quad n \in \mathbb{N}_0, \end{aligned} \tag{1.35}$$

$\{m_n^{(b)}(t, B(t))\}_{n\in\mathbb{N}_0}$ are martingales with respect to the filtration $\mathscr{F}_t$ generated by $B(t),\ t \geq 0$ as (1.6). The functions $\{m_n^{(b)}(t,x)\}_{n\in\mathbb{N}_0}$ are drift transformations of the fundamental martingale polynomials $\{m_n(t,x)\}_{n\in\mathbb{N}_0}$ associated with BM.

1.4 Quadratic Variation

Let $X(t),\ t \geq 0$ be a one-dimensional diffusion process on the probability space $(\Omega_X, \mathscr{F}_X, \mathrm{P}_X)$, where the expectation is written as E_X and the filtration is given by the *natural filtration* of X: $(\mathscr{F}_X)_t = \sigma(X(s) : 0 \leq s \leq t),\ t \geq 0$. For each time interval $[0,t], t > 0$, put $n \in \mathbb{N}$ and let $\Delta_n = \Delta_n([0,t])$ be a subdivision of $[0,t]$ with $0 \equiv t_0 < t_1 < \cdots < t_{n-1} < t_n \equiv t$. Then we define $Q^{\Delta_n}(t) = \sum_{m=1}^n (X(t_m) - X(t_{m-1}))^2$. If there is a process $Q(t),\ t \geq 0$ such that

$$\lim_{n\to\infty} \mathrm{P}_X[|Q^{\Delta_n}(t) - Q(t)| > \varepsilon] = 0, \quad {}^\forall \varepsilon > 0 \tag{1.36}$$

holds provided $\max_{1\leq m\leq n} |t_m - t_{m-1}| \to 0$ as $n \to \infty$, then we call $Q(t), t \geq 0$, the *quadratic variation* of $X(t),\ t \geq 0$ and express it by $\langle X, X\rangle_t,\ t \geq 0$.

For BM, $B(t),\ t \geq 0$, set $n \in \mathbb{N}$, $0 \equiv t_0 < t_1 < \cdots < t_{n-1} < t_n \equiv t$ and put $Q_{\mathrm{BM}}^{\Delta_n}(t) = \sum_{m=1}^n (B(t_m) - B(t_{m-1}))^2$. By the property **(BM3)**, the mean is given by $\mathrm{E}[Q_{\mathrm{BM}}^{\Delta_n}(t)] = \sum_{m=1}^n (t_m - t_{m-1}) = t$. The variance of $Q_{\mathrm{BM}}^{\Delta_n}(t)$

$$\begin{aligned} \sigma_{\mathrm{BM}}^{\Delta_n}(t)^2 &\equiv \mathrm{E}[(Q_{\mathrm{BM}}^{\Delta_n}(t) - t)^2] \\ &= \mathrm{E}\left[\left\{\sum_{m=1}^n \Big\{(B(t_m) - B(t_{m-1}))^2 - (t_m - t_{m-1})\Big\}\right\}^2\right] \end{aligned}$$

is calculated as

$$\begin{aligned} &\sum_{m=1}^n \Big\{\mathrm{E}[(B(t_m) - B(t_{m-1}))^4] - 2(t_m - t_{m-1})\mathrm{E}[(B(t_m) - B(t_{m-1}))^2] \\ &\qquad + (t_m - t_{m-1})^2\Big\} \\ &= \sum_{m=1}^n \Big\{3(t_m - t_{m-1})^2 - 2(t_m - t_{m-1})^2 + (t_m - t_{m-1})^2\Big\} \\ &= 2\sum_{m=1}^n (t_m - t_{m-1})^2 \leq 2t \max_{1\leq m\leq n} |t_m - t_{m-1}|, \end{aligned} \tag{1.37}$$

where independence of increments of BM mentioned in **(BM3)** and (1.124) in Exercise 1.5 were used. By *Chebyshev's inequality* (see Exercise 1.6), we have

$$\mathrm{P}[|Q^{\Delta_n}_{\mathrm{BM}}(t) - t| > \varepsilon] \le \frac{\sigma^{\Delta_n}_{\mathrm{BM}}(t)^2}{\varepsilon^2}, \quad {}^{\forall}\varepsilon > 0.$$

Provided that $\max_{1 \le m \le n} |t_m - t_{m-1}| \to 0$ as $n \to \infty$, (1.37) gives $\lim_{n\to\infty} \sigma^{\Delta_n}_{\mathrm{BM}}(t) = 0$ and it proves

$$\langle B, B\rangle_t = t, \qquad t \ge 0. \tag{1.38}$$

For a stopping time τ, we put $X^{\tau}(t) \equiv X(t \wedge \tau),\ t \ge 0$, where $t \wedge \tau \equiv \min\{t, \tau\}$. We define a diffusion process $X(t),\ t \ge 0$ as a *local martingale*, if there exists stopping times $\tau_n, n \in \mathbb{N}$ such that (i) the sequence $\{\tau_n\}_{n\in\mathbb{N}}$ is nondecreasing and $\lim_{n\to\infty} \tau_n = \infty$ a.s., and (ii) for every n, the process $X^{\tau_n}(t),\ t \ge 0$ is a martingale. When $X(t)$ is a local martingale, we can prove that a unique increasing continuous process is given by $\langle X, X\rangle_t,\ t \ge 0$ such that $\langle X, X\rangle_0 = 0$ and $X(t)^2 - \langle X, X\rangle_t,\ t \ge 0$ provides a local martingale.[1]

Assume that $X(t)$ and $Y(t),\ t \ge 0$ are both local martingales. Then $(X(t) + Y(t))^2 - \langle X + Y, X + Y\rangle_t$ and $(X(t) - Y(t))^2 - \langle X - Y, X - Y\rangle_t, t \ge 0$ are local martingales. Therefore, their difference $4X(t)Y(t) - \{\langle X+Y, X+Y\rangle_t - \langle X-Y, X-Y\rangle_t\},\ t \ge 0$ is also a local martingale. For any pair of local martingales $X(t)$ and $Y(t),\ t \ge 0$, we define the *mutual quadratic variation* (*cross variation*) *process* as

$$\langle X, Y\rangle_t \equiv \frac{1}{4}\{\langle X + Y, X + Y\rangle_t - \langle X - Y, X - Y\rangle_t\}, \quad t \ge 0. \tag{1.39}$$

We can prove that (Exercise 1.7), if $B_i(t),\ t \ge 0,\ i = 1, 2, \dots, D$ are independent BMs, then

$$\langle B_i, B_j\rangle_t = \delta_{ij} t, \quad 1 \le i, j \le D, \quad t \ge 0. \tag{1.40}$$

For a continuous process $A(t),\ t \ge 0$, here we consider the quantity

$$S^{\Delta_n}(t) = \sum_{m=1}^{n} |A(t_m) - A(t_{m-1})|$$

instead of $Q^{\Delta_n}(t)$, where $\Delta_n, n \in \mathbb{N}$ is a subdivision of the time interval $[0, t]$. We can see that if $\Delta_n \subset \Delta_{n+1}$, then $S^{\Delta_n}(t) \le S^{\Delta_{n+1}}(t), {}^{\forall}t \ge 0$. Assume that $\max_{1\le m\le n} |t_m - t_{m-1}| \to 0$ as $n \to \infty$. Let $\lim_{n\to\infty} \sup_{\Delta_n} S^{\Delta_n}(t) = S(t) \le \infty$

[1] Every martingale is a local martingale, but the converse is not true. If for every $a > 0$, the process $X^{\tau}(t),\ t \ge 0$, where τ ranges through all stopping times less than a with probability 1, is uniformly integrable, then the process is said of *class DL*. It is proved that a local martingale is also a martingale if and only if it is of class DL (see, for instance, Definition 4.8 in Chap. 1.4 of [3] and Proposition 1.7 in Chapter IV of [7]). BM is of class DL. We have already proved that $\langle B, B\rangle_t = t$ and $B(t)^2 - t$ is a martingale.

and call it the *variation* of A on $[0, t]$. If $S(t) < \infty$ for every t, then the process $A(t),\ t \geq 0$ is of *finite variation*. Let $S_{\rm BM}(t),\ t \geq 0$ be the variation of BM. We have

$$Q_{\rm BM}^{\Delta_n}(t) \leq \sup_{1 \leq m \leq n} |B(t_m) - B(t_{m-1})| S_{\rm BM}(t), \quad t \geq 0, \quad n \in \mathbb{N}.$$

By the property **(BM2)**, the RHS becomes 0 a.s. as $n \to \infty$ if $S_{\rm BM}(t)$ is finite. On the other hand, we have proved $Q_{\rm BM}^{\Delta_n}(t) \to \langle B, B \rangle_t$ as $n \to \infty$ in probability and the fact (1.38). Hence $S_{\rm BM}(t) = \infty$ a.s. $^\forall t > 0$.

1.5 Stochastic Integration

Assume that there is a strictly increasing sequence of times $\{t_m\}_{m=0}^{\infty}$ with $t_0 = 0$ and $\lim_{m \to \infty} t_m = \infty$. Let $\{\zeta_m(\omega)\}_{m=0}^{\infty}$ be a sequence of random variables such that $\sup_{m \geq 0} |\zeta_m(\omega)| \leq C$ with a nonrandom constant $C < \infty$ for every $\omega \in \Omega$, and ζ_m is $\mathscr{F}_{t_m}$-measurable for every $m \geq 0$. Let $\mathbf{1}_{(\omega)}$ be an *indicator function* for a condition (or an event) ω defined by

$$\mathbf{1}_{(\omega)} = \begin{cases} 1 \text{ if } \omega \text{ is satisfied (occurs),} \\ 0 \text{ otherwise.} \end{cases} \tag{1.41}$$

We consider a process given by

$$X(t, \omega) = \zeta_0(\omega)\mathbf{1}_{(t=0)} + \sum_{m=0}^{\infty} \zeta_m(\omega)\mathbf{1}_{(t \in (t_m, t_{m+1}])}, \quad t \geq 0, \quad \omega \in \Omega. \tag{1.42}$$

A process given in the form (1.42) is called *simple* and the class of all *simple processes* is denoted by $\mathscr{L}_0$. By definition, every sample path of X is left-continuous.

For such a simple process $X = \{X(t) : t \geq 0\}$, we consider

$$\begin{aligned} I[X](t) &\equiv \sum_{m=0}^{n-1} \zeta_m(\omega)(B(t_{m+1}, \omega) - B(t_m, \omega)) + \zeta_n(\omega)(B(t, \omega) - B(t_n, \omega)) \\ &= \sum_{m=0}^{\infty} \zeta_m(B(t \wedge t_{m+1}) - B(t \wedge t_m)), \quad t \geq 0, \end{aligned} \tag{1.43}$$

where n in the first line is the unique integer such that $t_n \leq t < t_{n+1}$, and we omit ω in the second line and below. It is obvious from the definition, $I[X](t),\ t \geq 0$ is a martingale. That is, I transforms a given simple process X to the unique martingale. We write this martingale transformation as

$$I[X](t) = \int_0^t X(s)dB(s) \tag{1.44}$$

and call it a *stochastic integral of X with respect to B*.

Assume that $t_n \le s < t_{n+1} \le t_\ell \le t < t_{\ell+1}$ and consider

$$\mathrm{E}[\{I[X](t) - I[X](s)\}^2|\mathscr{F}_s] = \mathrm{E}\bigg[\bigg\{\zeta_n(B(t_{n+1}) - B(s))$$
$$+ \sum_{m=n+1}^{\ell-1} \zeta_m(B(t_{m+1}) - B(t_m)) + \zeta_\ell(B(t) - B(t_\ell))\bigg\}^2\bigg|\mathscr{F}_s\bigg]. \tag{1.45}$$

By the property **(BM3)** of BM, the increments of BM in different time intervals are independent from each other and have mean 0. Then the above is equal to

$$\mathrm{E}[\zeta_n^2(B(t_{n+1}) - B(s))^2|\mathscr{F}_s] + \sum_{m=n+1}^{\ell-1} \mathrm{E}[\zeta_m^2(B(t_{m+1}) - B(t_m))^2|\mathscr{F}_s]$$
$$+ \mathrm{E}[\zeta_\ell^2(B(t) - B(t_\ell))^2|\mathscr{F}_s].$$

Since the variance of BM in each time interval is equal to the time duration of interval by the property **(BM3)**, this is calculated as

$$\mathrm{E}[\zeta_n^2|\mathscr{F}_s](t_{n+1} - s) + \sum_{m=n+1}^{\ell-1} \mathrm{E}[\zeta_m^2|\mathscr{F}_s](t_{m+1} - t_m) + \mathrm{E}[\zeta_\ell^2|\mathscr{F}_s](t - t_\ell)$$
$$= \mathrm{E}\left[\zeta_n^2(t_{n+1} - s) + \sum_{m=n+1}^{\ell-1} \zeta_m^2(t_{m+1} - t_m) + \zeta_\ell^2(t - t_\ell)\middle|\mathscr{F}_s\right].$$

It defines $\mathrm{E}\left[\int_s^t X(r)^2dr\middle|\mathscr{F}_s\right]$ and thus (1.45) is equal to

$$\mathrm{E}\left[\int_0^t X(r)^2dr - \int_0^s X(r)^2dr\middle|\mathscr{F}_s\right] = \mathrm{E}\left[\int_0^t X(r)^2dr\middle|\mathscr{F}_s\right] - \int_0^s X(r)^2dr, \quad \text{a.s.}$$

On the other hand, the LHS of (1.45) is written as

$$\mathrm{E}\Big[\{I[X](t) - I[X](s)\}^2\Big|\mathscr{F}_s\Big]$$
$$= \mathrm{E}\Big[I[X](t)^2 - I[X](s)^2 - 2(I[X](t) - I[X](s))I[X](s)\Big|\mathscr{F}_s\Big]$$
$$= \mathrm{E}\Big[I[X](t)^2 - I[X](s)^2\Big|\mathscr{F}_s\Big] = \mathrm{E}[I[X](t)^2|\mathscr{F}_s] - I[X](s)^2, \quad \text{a.s.}$$

Then, we have the equality

$$\mathrm{E}\left[I[X](t)^2 - \int_0^t X(r)^2dr\middle|\mathscr{F}_s\right] = I[X](s)^2 - \int_0^s X(r)^2dr, \quad \text{a.s.,}$$

that is, $I[X](t)^2 - \int_0^t X(r)^2 dr,\ t \geq 0$ is a martingale. Since $I[X](t)$ is a martingale, its quadratic variation $\langle I[X], I[X]\rangle_t,\ t \geq 0$ is the unique increasing continuous process such that $\langle I[X], I[X]\rangle_0 = 0$ and $I[X](t)^2 - \langle I[X], I[X]\rangle_t,\ t \geq 0$ gives a local martingale. Therefore, we can conclude that for the stochastic integral (1.44) of the simple process (1.42), its quadratic variation is given by

$$\langle I[X], I[X]\rangle_t = \int_0^t X(s)^2 ds, \quad t \geq 0.$$

The above results for the BM and $X \in \mathscr{L}_0$ can be generalized as follows [3, 7]. The stochastic process $X = \{X(t) : t \geq 0\}$ is said to be *adapted* to the filtration $\{\mathscr{F}_t : t \geq 0\}$ if, for each $t \geq 0$, $X(t)$ is an $\mathscr{F}_t$-measurable random variable. If X is adapted to $\{\mathscr{F}_t : t \geq 0\}$ and every sample path is left-continuous (or right-continuous), then X can be said to be *progressively measurable* with respect to $\{\mathscr{F}_t : t \geq 0\}$. Let $\mathscr{L}^*$ denote the set of all progressively measurable processes satisfying

$$\int_0^T X(t)^2 dt < \infty, \quad {}^\forall T > 0.$$

If a martingale $M = \{M(t) : t \geq 0\}$ satisfies $\mathrm{E}[M(t)^2] < \infty$ for every $t \geq 0$, it is said to be *square-integrable*. The set of all square-integrable and continuous martingales is denoted by $\mathscr{M}_2^c$. For $M \in \mathscr{M}_2^c$ and $X \in \mathscr{L}^*$, the stochastic integral of X with respect to M

$$I[X](t) = \int_0^t X(s) dM(s)$$

can be defined and we have the following properties:

$$I[X](0) = 0, \tag{1.46}$$

$$\mathrm{E}[I[X](t) | \mathscr{F}_s] = I[X](s), \quad 0 \leq s < t < \infty, \tag{1.47}$$

$$\langle I[X], I[X]\rangle_t = \int_0^t X(s)^2 d\langle M, M\rangle_s, \quad t \geq 0, \tag{1.48}$$

where $d\langle M, M\rangle_t = \langle dM, dM\rangle_t$, and

$$I[\alpha X + \beta Y](t) = \alpha I[X](t) + \beta I[Y](t), \quad t \geq 0, \tag{1.49}$$

for $\alpha, \beta \in \mathbb{C},\ X, Y \in \mathscr{L}^*$.

Moreover, if $M, N \in \mathscr{M}_2^c$ and $X, Y \in \mathscr{L}^*$, the mutual quadratic variation of

$$I[X] = \int_0^t X(s) dM(s) \quad \text{and} \quad J[Y] = \int_0^t Y(s) dN(s)$$

is given by

$$\langle I[X], J[Y]\rangle_t = \int_0^t X(s)Y(s)d\langle M, N\rangle_s, \quad t \geq 0. \tag{1.50}$$

The differential form is

$$d\langle I[X], J[Y]\rangle_t = X(t)Y(t)d\langle M, N\rangle_t, \quad t \geq 0. \tag{1.51}$$

1.6 Itô's Formula

Let $N \in \mathbb{N}$ and $\{X_1(t), \ldots, X_N(t)\}$, $t \geq 0$ be a set of diffusion processes. Put $\mathbf{X}(t) = (X_1(t), \ldots, X_N(t))$. Let F be a real function of $(t, \mathbf{x}) \in [0, \infty) \times \mathbb{R}^N$, which is bounded and has a bounded first-order derivative with respect to t and bounded first- and second-order derivatives with respect to x_j, $1 \leq j \leq N$, and we denote this by $F \in \mathrm{C}_{\mathrm{b}}^{1,2}$. We know that, in order to describe the statistics of a function of several random variables, we have to take into account the 'propagation of error'. For the process $F(t, \mathbf{X}(t))$ that is defined as a function of t as well as a functional of processes $X_1(t), \ldots, X_N(t)$, $t \geq 0$, *Itô's formula* gives an equation which governs the differential of $F(t, \mathbf{X}(\cdot))$ as

$$\begin{aligned} dF(t, \mathbf{X}(t)) &= \sum_{i=1}^{N} \frac{\partial F}{\partial x_i}(t, \mathbf{X}(t))dX_i(t) + \frac{\partial F}{\partial t}(t, \mathbf{X}(t))dt \\ &\quad + \frac{1}{2} \sum_{1 \leq i,j \leq N} \frac{\partial^2 F}{\partial x_i \partial x_j}(t, \mathbf{X}(t))d\langle X_i, X_j\rangle_t, \quad t \geq 0. \end{aligned} \tag{1.52}$$

The integral form of Itô's formula is expressed by

$$\begin{aligned} F(t, \mathbf{X}(t)) &= F(0, \mathbf{X}(0)) + \sum_{i=1}^{N} \int_0^t \frac{\partial F}{\partial x_i}(s, \mathbf{X}(s))dX_i(s) + \int_0^t \frac{\partial F}{\partial t}(s, \mathbf{X}(s))ds \\ &\quad + \frac{1}{2} \sum_{1 \leq i,j \leq N} \int_0^t \frac{\partial^2 F}{\partial x_i \partial x_j}(s, \mathbf{X}(s))d\langle X_i, X_j\rangle_s, \quad t \geq 0. \end{aligned} \tag{1.53}$$

A continuous process X given by the sum of a local martingale M and a finite-variation process A, $X(t) = M(t) + A(t)$, $t \geq 0$, is called a *semimartingale*. When $X_i(t) = M_i(t) + A_i(t), t \geq 0, 1 \leq i \leq N$, are semimartingales, the local martingale part of $F(t, \mathbf{X}(t)), t \geq 0$, is given by $\sum_{i=1}^N \int_0^t \partial F/\partial x_i(s, \mathbf{X}(s))dM_i(s)$, $t \geq 0$, which is derived from the second term in the RHS of (1.53). Other terms in the RHS of (1.53) including $\sum_{i=1}^N \int_0^t \partial F/\partial x_i(s, \mathbf{X}(s))dA_i(s)$, $t \geq 0$, give the finite-variation part for $F(t, \mathbf{X}(t)), t \geq 0$. We will use the fact that if the continuous process $F(t, \mathbf{X}(t))$, $t \geq 0$ is a local martingale, then its finite-variation part should vanish, and vice versa. (See Exercise 1.8.)

1.7 Complex Brownian Motion and Conformal Invariance

The *complex Brownian motion* is defined by

$$Z(t) = B(t) + \sqrt{-1}\widetilde{B}(t), \quad t \geq 0. \tag{1.54}$$

Its probability space is a product of the space $(\Omega, \mathscr{F}, \mathrm{P})$ for B and the space $(\widetilde{\Omega}, \widetilde{\mathscr{F}}, \widetilde{\mathrm{P}})$ for $\widetilde{B}$, and it is written as $(\Omega_Z, \mathscr{F}_Z, \mathrm{P}_Z)$ with expectation E_Z. Both of the real and imaginary parts are real martingales and so $Z(t),\ t \geq 0$ is a complex martingale. Since $\langle B, B\rangle_t = \langle \widetilde{B}, \widetilde{B}\rangle_t = t$ and $\langle B, \widetilde{B}\rangle_t = 0$ by (1.40), we see that

$$\begin{aligned}\langle Z, Z\rangle_t &= \langle B + \sqrt{-1}\widetilde{B}, B + \sqrt{-1}\widetilde{B}\rangle_t \\ &= \langle B, B\rangle_t - \langle \widetilde{B}, \widetilde{B}\rangle_t + 2\sqrt{-1}\langle B, \widetilde{B}\rangle_t = 0.\end{aligned}$$

It implies that $Z(t)^2,\ t \geq 0$ is a martingale.

If F is a complex function of $z = x + \sqrt{-1}y$ with $x, y \in \mathbb{R}$, which is in C_b^2 as a function of x and y and it does not depend on t explicitly, then Itô's formula (1.52) gives

$$dF(Z(t)) = \frac{\partial F}{\partial z}(Z(t))dZ(t) + \frac{\partial F}{\partial \overline{z}}(Z(t))d\overline{Z}(t) + \frac{1}{4}\Delta F(Z(t))d\langle Z, \overline{Z}\rangle_t, \tag{1.55}$$

where $\overline{z} = x - \sqrt{-1}y$ denotes the complex conjugate of $z = x + \sqrt{-1}y,\ x, y \in \mathbb{R}$, and $\Delta = \partial^2/\partial x^2 + \partial^2/\partial y^2 = 4\partial^2/\partial z \partial \overline{z}$. Therefore, if F is harmonic (i.e., $\Delta F(z) = 0$), $F(Z(t)),\ t \geq 0$ is a local martingale, and if F is holomorphic (i.e., a function of z but not of $\overline{z}$), then

$$F(Z(t)) = F(Z(0)) + \int_0^t F'(Z(s))dZ(s), \quad t \geq 0.$$

Moreover, we can prove that, if F is an entire and non-constant function, then $F(Z(t)), t \geq 0$ is a time change of a complex Brownian motion. That is, if we set $X(t) = \Re F(Z(t)), Y(t) = \Im F(Z(t)), t \geq 0$, then there is a complex Brownian motion $\widehat{Z}$ in $(\Omega_Z, \mathscr{F}_Z, \mathrm{P}_Z)$ such that

$$F(Z(t)) = F(Z(0)) + \widehat{Z}(\langle X, X\rangle_t), \quad t \geq 0,$$

where $\langle X, X\rangle_t = \int_0^t |F'(Z(s))|^2 ds = \langle Y, Y\rangle_t$ is a strictly increasing function and $\langle X, X\rangle_\infty = \infty$. This result is known as *the conformal invariance of complex Brownian motion.*

$F(z) = z^n, n \in \mathbb{N}_0$ are entire and non-constant functions. Then $(Z(t))^n,\ n \in \mathbb{N}_0$ are time-changed complex Brownian motions and thus martingales:

$$\mathrm{E}_Z[Z(t)^n | (\mathscr{F}_Z)_s] = Z(s)^n, \quad 0 \leq s \leq t, \quad n \in \mathbb{N}_0, \quad \text{a.s.}$$

By taking the expectation with respect to $\widetilde{B}(\cdot)$ of both sides in this equation, we obtain

$$\mathrm{E}[\widetilde{\mathrm{E}}[(B(t)+\sqrt{-1}\widetilde{B}(t))^n]|\mathscr{F}_s] = \widetilde{\mathrm{E}}[(B(s)+\sqrt{-1}\widetilde{B}(s))^n],$$

which gives (1.21) through (1.27).

1.8 Stochastic Differential Equations for Bessel Processes

Let $D \in \mathbb{N}$ denote the spatial dimension. For $D \geq 2$, the D-dimensional BM in $\mathbb{R}^D$ starting from the position $\mathbf{x} = (x_1, \dots, x_D) \in \mathbb{R}^D$ is defined by the D-dimensional vector-valued diffusion process,

$$\mathbf{B}^{\mathbf{x}}(t) = (B_1^{x_1}(t), B_2^{x_2}(t), \dots, B_D^{x_D}(t)), \quad t \geq 0, \tag{1.56}$$

where $\{B_i^{x_i}(t)\}_{i=1}^D$, $t \geq 0$ are independent one-dimensional BMs.

The D-dimensional *Bessel process* is defined as the absolute value (i.e., the radial coordinate) of the D-dimensional Brownian motion,

$$R^x(t) \equiv |\mathbf{B}^{\mathbf{x}}(t)| = \sqrt{B_1^{x_1}(t)^2 + \cdots + B_D^{x_D}(t)^2}, \quad t \geq 0, \tag{1.57}$$

where the initial value is given by $R^x(0) = x = |\mathbf{x}| = \sqrt{x_1^2 + \cdots + x_D^2} \geq 0$. See Fig. 1.1. By definition $R^x(t)$ is nonnegative, $R^x(t) \in \mathbb{R}_+ \cup \{0\}$, where $\mathbb{R}_+ \equiv \{x \in \mathbb{R} : x > 0\}$. We will abbreviate the D-dimensional Bessel process to BES$^{(D)}$.

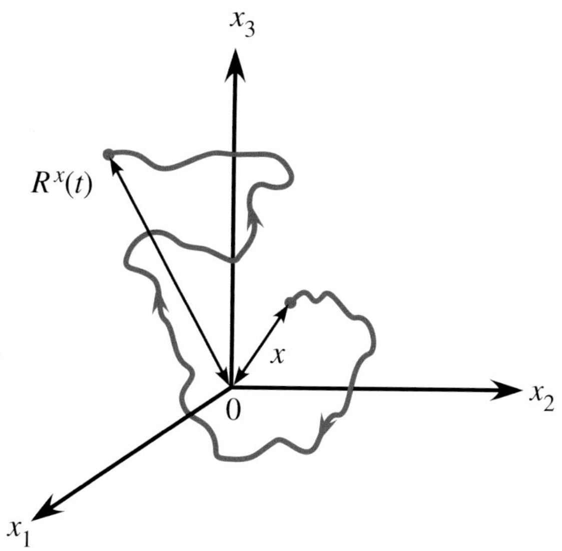

Fig. 1.1 The D-dimensional Bessel process $R^x(t)$ is defined as the radial part of the BM in the D-dimensional space. The initial value x of the Bessel process is the distance between the origin and the position where the BM is started. The figure shows the case where $D = 3$

By this definition, $R^x(t)$ is a functional of D-tuples of diffusion processes $\{B_i^{x_i}(t)\}_{i=1}^D$, $t \geq 0$. Now we apply Itô's formula (1.52) to $\text{BES}^{(D)}$ (1.57). Assume $x = |\mathbf{x}| > 0$. In this case

$$R^x(t) = F(\mathbf{B}(t)), \quad t \geq 0 \quad \text{with} \quad F(\mathbf{y}) = \sqrt{\sum_{i=1}^D y_i^2}. \tag{1.58}$$

We see that

$$\frac{\partial F}{\partial t} = 0, \quad \frac{\partial F}{\partial y_i} = \frac{y_i}{F}, \quad \frac{\partial^2 F}{\partial y_i \partial y_j} = \frac{\delta_{ij}}{F} - \frac{y_i y_j}{F^3}, \quad 1 \leq i, j \leq D.$$

From (1.40), we have $d\langle B_i, B_j \rangle_t = \langle dB_i, dB_j \rangle_t = \delta_{ij} dt$, $1 \leq i, j \leq D$, $t \geq 0$. Then, the third term of (1.52) for $\text{BES}^{(D)}$ becomes

$$\frac{1}{2} \sum_{1 \leq i,j \leq D} \left\{ \frac{\delta_{ij}}{F(\mathbf{B}(t))} - \frac{B_i(t) B_j(t)}{F(\mathbf{B}(t))^3} \right\} \delta_{ij} dt = \frac{D-1}{2} \frac{1}{F(\mathbf{B}(t))} dt = \frac{D-1}{2} \frac{dt}{R^x(t)}.$$

On the other hand, the first term of (1.52) for $\text{BES}^{(D)}$ is

$$\frac{1}{R^x(t)} \sum_{i=1}^D B_i(t) dB_i(t). \tag{1.59}$$

It seems to be complicated, but (1.51) enables us to calculate its quadratic variation as

$$\begin{aligned} \left\langle \frac{1}{R^x} \sum_{i=1}^D B_i dB_i, \frac{1}{R^x} \sum_{j=1}^D B_j dB_j \right\rangle_t &= \frac{1}{R^x(t)^2} \sum_{i=1}^D \sum_{j=1}^D B_i(t) B_j(t) \langle dB_i, dB_j \rangle_t \\ &= \frac{1}{R^x(t)^2} \sum_{i=1}^D \sum_{j=1}^D B_i(t) B_j(t) \delta_{ij} dt = dt, \end{aligned}$$

where the independence of BMs (1.40) and the definition, $R^x(t)^2 = \sum_{i=1}^D B_i(t)^2$, have been used. That is, (1.59) is equivalent in probability law with an infinitesimal increment of a diffusion process with quadratic variation dt. Then, by introducing a BM, $B^x(t)$, $t \geq 0$, which is different from $B_i^{x_i}(t)$, $t \geq 0$, $x_i \in \mathbb{R}$, $i = 1, 2, \ldots, D$ and is started at $x = |\mathbf{x}| > 0$, (1.59) is identified with $dB^x(t)$, $t \geq 0$. We have thus obtained the following equation for $\text{BES}^{(D)}$,

$$dR^x(t) = dB^x(t) + \frac{D-1}{2} \frac{dt}{R^x(t)}, \quad x > 0, \quad 0 \leq t < T^x, \tag{1.60}$$

where $T^x = \inf\{t > 0 : R^x(t) = 0\}$.

The first term of the RHS, $dB^x(t)$, denotes the infinitesimal increment of BM starting from $x > 0$ at time $t = 0$. This martingale term gives randomness to the motion. On the other hand, if $D > 1$, for $dt > 0$, the second term in the RHS of (1.60) is positive definite. It means that there is a drift to increase the value of $R^x(t)$. This drift term is increasing in D and decreasing in $R^x(t)$. Since as $R^x(t) \downarrow 0$, the drift term $\uparrow \infty$, it seems that a 'repulsive force' is acting to the D-dimensional BM, $\mathbf{B}^{\mathbf{x}}(t)$, $|\mathbf{x}| > 0$ to keep the distance from the origin be positive, $R^x(t) = |\mathbf{B}^{\mathbf{x}}(t)| > 0$ and avoid a collision of the Brownian particle with the origin. A differential equation such as (1.60), which involves a random fluctuation term and a drift term is called a *stochastic differential equation* (SDE). The integral form is written as

$$R^x(t) = x + B(t) + \frac{D-1}{2} \int_0^t \frac{ds}{R^x(s)}, \quad x > 0, \quad 0 \le t < T^x. \tag{1.61}$$

What is the origin of the repulsive force between the D-dimensional BM and the origin? Why does $\mathbf{B}^{\mathbf{x}}(t)$ starting from a point $\mathbf{x} \neq 0$ not want to return to the origin? Why is the strength of the outward drift increasing in the dimension $D > 1$?

There is no positive reason for $\mathbf{B}^{\mathbf{x}}(t)$ to avoid visiting the origin, since by the definition (1.56) all components $B_i^{x_i}(t)$, $1 \le i \le N$ enjoy independent BMs. As the dimension of space D increases, however, the possibility *not* to visit the origin (or any specified point) increases, since among D directions in the space only one direction is toward the origin (or toward the specified point) and other $D-1$ directions are orthogonal to it. If one knows the second law of thermodynamics, which is also called the *law of increasing entropy*, one will understand that we would like to say here that the repulsive force acting from the origin to the Bessel process is not a usual force treated in mechanics but an 'entropy force'. (Note that the physical dimension of entropy [J/K] is different from that of force in mechanics [N] $\equiv$ [J/m].) Anyway, the important fact is that, while the fluctuation (quadratic variation) of the BM is given as $\langle B, B\rangle_t = t$, $t \ge 0$, independently of D, the strength of repulsive drift is increasing in D. Then, the return probability of $R^x(t)$, $x > 0$ to the origin should be a nonincreasing function of D.

The following equivalence in probability law is established for arbitrary $x > 0$,

$$\left(\frac{1}{x} R^x(x^2 t)\right)_{t \ge 0} \stackrel{\text{(law)}}{=} (R^1(t))_{t \ge 0}. \tag{1.62}$$

It inherits (1.4) and is called the *scaling property of the Bessel process* (Exercises 1.9 and 1.10).

1.9 Kolmogorov Equation

Let $X(t)$, $t \ge 0$ be a one-dimensional diffusion in $(\Omega_X, \mathscr{F}_X, \mathrm{P}_X)$, which satisfies the following SDE:

$$dX(t) = \sigma(X(t))dB(t) + b(X(t))dt, \quad X(0) = x. \tag{1.63}$$

Here $B(t),\ t \geq 0$ is BM and the functions $\sigma, b : \mathbb{R} \mapsto \mathbb{R}$ satisfy the condition that $^{\exists}K \geq 0$, s.t. $|\sigma(x) - \sigma(y)| \leq K|x - y|$, $|b(x) - b(y)| \leq K|x - y|$, $x, y \in \mathbb{R}$. (This is called the *Lipschitz continuity*.) Put

$$u(s, x) = \mathrm{E}^x[f(X(T - s))], \quad 0 \leq s < T < \infty, \tag{1.64}$$

with an $(\mathscr{F}_X)_T$-measurable bounded function f. By the Markov property (1.8), for $0 \leq s < t < T < \infty$,

$$\begin{aligned} u(s, x) &= \mathrm{E}^x\left[\mathrm{E}^{X(t-s)}[f(X(T - t))]\right] \\ &= \mathrm{E}^x[u(t, X(t - s))]. \end{aligned} \tag{1.65}$$

Assume that $u(s, x) \in \mathrm{C}_{\mathrm{b}}^{1,2}$, i.e., bounded and having bounded first-order derivative with respect to time s and bounded first-order and second-order derivatives with respect to space x. Then by Itô's formula (1.52),

$$\begin{aligned} u(t, X(t - s)) - u(s, x) = &\int_0^{t-s} \left(\frac{\partial u}{\partial s} + \mathsf{L}u\right)(s + r, X(r))dr \\ &+ \int_0^{t-s} \sigma(X(r))\frac{\partial u}{\partial x}(s + r, X(r))dB(r), \end{aligned} \tag{1.66}$$

where

$$(\mathsf{L}f)(x) = \frac{1}{2}a(x)\frac{d^2}{dx^2}f(x) + b(x)\frac{d}{dx}f(x) \tag{1.67}$$

with

$$a(x) = \sigma(x)^2. \tag{1.68}$$

Since (1.65) holds, taking expectation of (1.66) gives

$$\mathrm{E}^x\left[\int_0^{t-s} \left(\frac{\partial u}{\partial s} + \mathsf{L}u\right)(s + r, X(r))dr\right] = 0,$$

where the expectation of the second term in the RHS of (1.66) vanished, for it is a martingale. Set $h = t - s$, divide both sides of the above equation by h and take the limit $h \to 0$. Then we have

$$\frac{\partial u(s, x)}{\partial s} + \mathsf{L}u(s, x) = 0. \tag{1.69}$$

Since L acts as a differential operator with respect to the initial value x, it is called the *backward Kolmogorov equation*. The differential operator (1.67) is called the *generator* of the diffusion process.

The SDE for $\text{BES}^{(D)}$ is given by the equation (1.63) with $\sigma(x) \equiv 1$ and $b(x) = (D-1)/(2x)$. Then the generator of $\text{BES}^{(D)}$ is obtained as

$$\mathsf{L}^{(D)} = \frac{1}{2}\frac{\partial^2}{\partial x^2} + \frac{D-1}{2x}\frac{\partial}{\partial x}. \tag{1.70}$$

Let $p^{(D)}(t-s, y|x)$ be the transition probability density of $\text{BES}^{(D)}$ from x at times $s \geq 0$ to y at times $t \geq s$. For any $t \geq s,\ y \in \mathbb{R}_+,\ p^{(D)}(t-s, y|x),\ x > 0$ solves (1.69) with (1.70) under the condition $\lim_{s \uparrow t} p^{(D)}(t-s, y|x) = \delta(x-y)$. In other words, $p^{(D)}(t, y|x)$ solves

$$\frac{\partial}{\partial t} p^{(D)}(t, y|x) = \mathsf{L}^{(D)} p^{(D)}(t, y|x) \tag{1.71}$$

under the initial condition $p^{(D)}(0, y|x) = \delta(x-y)$, which is called the backward Kolmogorov equation for $\text{BES}^{(D)}$.

Let $I_\nu(z)$ be the *modified Bessel function* of the first kind defined by

$$I_\nu(z) = \sum_{n=0}^{\infty} \frac{1}{\Gamma(n+1)\Gamma(n+1+\nu)} \left(\frac{z}{2}\right)^{2n+\nu} \tag{1.72}$$

with the gamma function

$$\Gamma(z) = \int_0^\infty e^{-u} u^{z-1} du, \quad \Re z > 0. \tag{1.73}$$

The function $I_\nu(z)$ solves the *Bessel differential equation*

$$\frac{d^2 w}{dz^2} + \frac{1}{z}\frac{dw}{dz} - \left(1 + \frac{\nu^2}{z^2}\right) w = 0. \tag{1.74}$$

Then we can show that

$$p^{(D)}(t, y|x) = \begin{cases} \dfrac{1}{t}\dfrac{y^{\nu+1}}{x^\nu} e^{-(x^2+y^2)/2t} I_\nu\left(\dfrac{xy}{t}\right), & t > 0,\ x > 0,\ y \geq 0, \\ \dfrac{y^{2\nu+1}}{2^\nu t^{\nu+1} \Gamma(\nu+1)} e^{-y^2/2t}, & t > 0,\ x = 0,\ y \geq 0 \\ \delta(y-x), & t = 0,\ x, y \geq 0, \end{cases} \tag{1.75}$$

where the index ν is specified by the dimension D as

$$\nu = \frac{D-2}{2} \iff D = 2(\nu + 1). \tag{1.76}$$

This fact that $p^{(D)}(t, y|x)$ is expressed using $I_\nu(z)$ gives the reason why the process $R^x(t)$ is called the Bessel process (see Exercise 1.11).

1.10 BES$^{(3)}$ and Absorbing BM

When $D = 3, \nu = 1/2$ by (1.76), and we can use the equality $I_{1/2}(z) = \sqrt{2/\pi z} \sinh z = (e^z - e^{-z})/\sqrt{2\pi z}$. Then (1.75) gives

$$p^{(3)}(t, y|x) = \frac{y}{x}\Big\{p(t, y|x) - p(t, y| - x)\Big\} \tag{1.77}$$

for $t > 0, x > 0, y \geq 0$, where $p(t, y|x)$ is the transition probability density of BM started at x given by (1.1). If we put

$$q^{\text{abs}}(t, y|x) = p(t, y|x) - p(t, y| - x), \tag{1.78}$$

we see that $q^{\text{abs}}(t, 0|x) = 0$ for any $x > 0$, since the transition probability density of BM, $p(t, y|x)$, is an even function of $y - x$.

In Fig. 1.2a, one realization of a Brownian path from $x > 0$ to $y > 0$ is represented by a red curve and denoted by path A, which visits the nonpositive region $\mathbb{R}_- \cup \{0\}$, where $\mathbb{R}_- \equiv \{x \in \mathbb{R} : x < 0\}$. The first time the path A hits the origin is denoted by τ. Path B, which is represented by a black curve, is a mirror image of path A with respect to the origin $x = 0$, which is running from $-x < 0$ to $-y < 0$. Path C (blue

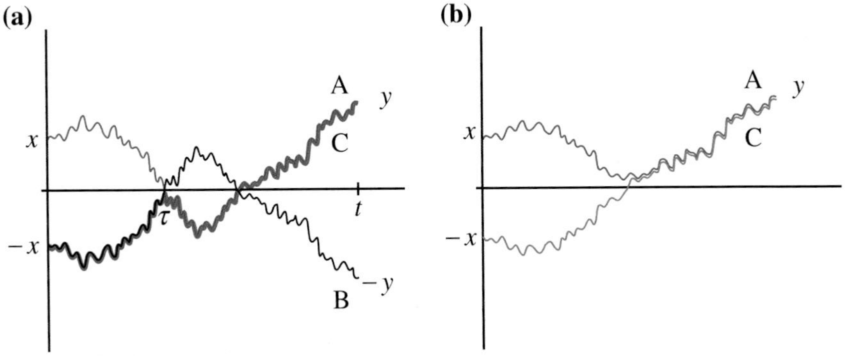

Fig. 1.2 Brownian path from $x > 0$ to $y > 0$ contributing to $p(t, y|x)$ and that from $-x < 0$ to $y > 0$ contributing to $p(t, y| - x)$

curve) is then defined as the concatenation of the part of path B from time zero up to time τ and the part of path A after τ such that it runs from $-x < 0$ to $y > 0$. By the reflection principle of BM (1.5) applied at time τ, we see that there exists a bijection between path A and path C, which have the same probability weight as Brownian paths. Since the Brownian path A contributes to $p(t, y|x)$ and the Brownian path C contributes to $p(t, y|-x)$, such a path from $x > 0$ to $y > 0$ that visits $\mathbb{R}_-$ is cancelled in $q^{\text{abs}}(t, y|x)$ given by (1.78). In Fig. 1.2b, a path from $x > 0$ to $y > 0$ which stays in the positive region $\mathbb{R}_+$ is considered (path A). In this case there is no path that perfectly cancels the contribution of path A to (1.78) as path B does in the case (a). In summary, $q^{\text{abs}}(t, y|x) = p(t, y|x) - p(t, y|-x)$ gives the total weight of Brownian paths which do not hit the origin.

We consider the situation where an absorbing wall is put at the origin and, if the Brownian particle starting from $x > 0$ arrives at the origin, it is absorbed there and the motion is stopped. Such a process is called the *absorbing Brownian motion in* $\mathbb{R}_+$. Its transition probability density is given by q^{abs}.

By absorption, the total mass of paths from $x > 0$ to $y > 0$ is then reduced, if we compare the original BM and the absorbing Brownian motion in $\mathbb{R}_+$. The factor y/x appearing in the transition probability density (1.77) of BES$^{(3)}$ is for renormalization so that $\int_{\mathbb{R}_+} p^{(3)}(t, y|x)dy = 1, {}^\forall t > 0, {}^\forall x > 0$ (see Exercises 1.12 and 1.13). We regard this renormalization procedure from q^{abs} to $p^{(3)}$ as a transformation. Since x is a one-dimensional harmonic function in a rather trivial sense $\Delta^{(1)}x \equiv d^2x/dx^2 = 0$, we say that the BES$^{(3)}$ is a *harmonic transformation* (h-transformation) of the one-dimensional absorbing BM in the sense of Doob [2]. This implies the following equivalence [5] (see Exercise 1.14).

$$\text{BES}^{(3)} \iff \text{one-dimensional Brownian motion conditioned to stay positive}$$

Let $\mathbb{E}^x_{\text{BES}^{(3)}}$ denote the expectation with respect to BES$^{(3)}$, $(R(t))_{t\geq 0}$, started at $x \in \mathbb{R}_+$. For an independent BM, $(B(t))_{t\geq 0}$ started at the same point $x \in \mathbb{R}_+$, let $\tau = \inf\{t > 0 : B(t) = 0\}$. Then the above equivalence is written as follows: for any $\mathscr{F}_t$-measurable bounded function F, $t \geq 0$,

$$\mathbb{E}^x_{\text{BES}^{(3)}}[F(R(t))] = \mathrm{E}^x\left[F(B(t))\mathbf{1}_{(\tau>t)}\frac{B(t)}{x}\right], \quad t \geq 0, \tag{1.79}$$

where E^x is an expectation with respect to BM started at $x \in \mathbb{R}_+$. If F is an even function: $F(-x) = F(x)$, the above gives

$$\mathbb{E}^x_{\text{BES}^{(3)}}[F(R(t))] = \mathrm{E}^x\left[F(B(t))\frac{B(t)}{x}\right], \quad t \geq 0, \tag{1.80}$$

since by the reflection principle of BM (1.5), all contribution from paths $\{\omega : \tau \leq t\}$ should be canceled out (see Exercise 1.15).

Here we emphasize the obvious fact that $p^{(3)}(t, 0|x) = 0, {}^{\forall} x > 0$. It implies that $\text{BES}^{(3)}$ does not visit the origin. When $D = 3$, the outward drift is strong enough to avoid any visit to the origin. Moreover, we can prove that for any $x > 0$, $R^x(t) \to \infty$ as $t \to \infty$ with probability 1 and we say the process is *transient* (see Theorem 1.1 (ii) below).

1.11 $\text{BES}^{(1)}$ and Reflecting BM

When $D = 1, \nu = -1/2$ by (1.76) and we use the equality $I_{-1/2}(z) = \sqrt{2/\pi z} \cosh z = (e^z + e^{-z})/\sqrt{2\pi z}$. In this case (1.75) gives

$$p^{(1)}(t, y|x) = p(t, y|x) + p(t, y| - x) \tag{1.81}$$

for $t > 0, x, y \geq 0$. As illustrated by Fig. 1.3, the one-dimensional BM visits the origin frequently (see Theorem 1.1 (iv) below). For a Brownian path starting from $x > 0$ represented by a red curve (path A), its mirror image with respect to the origin is represented by a blue curve (path B), which starts from $-x < 0$. If we observe the motion only in the nonnegative region $\mathbb{R}_+ \cup \{0\}$, the superposition of Brownian paths A and B gives the path of a *reflecting Brownian motion*, where a reflecting wall is put at the origin. (Note that the transition probability density $p^{(1)}$ is the duplicate of p as (1.81), but the space is halved as $x, y \in \mathbb{R} \mapsto x, y \in \mathbb{R}_+ \cup \{0\}$.) Then the equality (1.81) implies the following equivalence:

$$\text{BES}^{(1)} \iff \text{one-dimensional reflecting Brownian motion.}$$

This is of course a direct consequence of the definition of Bessel process (1.57), since it gives $R^x(t) = |B^x(t)|$ in $D = 1$.

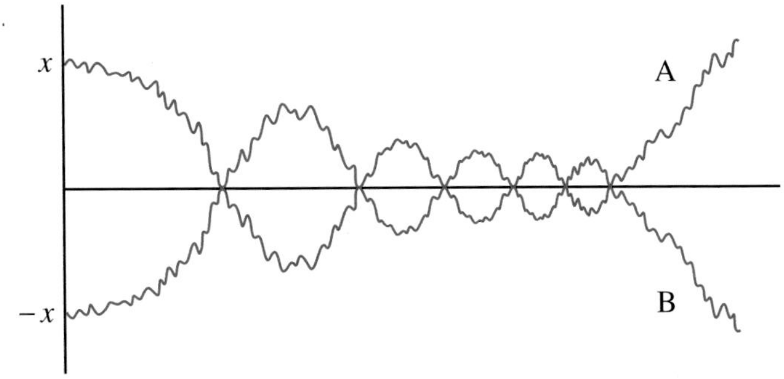

Fig. 1.3 A typical path of one-dimensional Brownian motion starting from $x > 0$ and its mirror image. They visit the origin frequently

The important fact is that the one-dimensional BM starting from $x \neq 0$ visits the origin frequently and we say that the one-dimensional Bessel process is *recurrent*. Remark that in Eqs. (1.60) and (1.71), the drift terms vanish when $D = 1$. So we have to assume the reflecting boundary condition at the origin when we discuss BES$^{(1)}$ instead of the one-dimensional BM.

1.12 Critical Dimension $D_c = 2$

Now the following question is addressed: At which dimension does the Bessel process change its property from being recurrent to transient?

Before answering this question, here we would like to extend the setting of the question. Originally, the Bessel process was defined by (1.57) for $D \in \mathbb{N}$. We find that, however, the modified Bessel function (1.72) is an analytic function of ν for all values of ν. So we will be able to define the Bessel process for any value of dimension $D \geq 1$ as a diffusion process in $\mathbb{R}_+$ such that the transition probability density function is given by (1.75), where the index $\nu \geq -1/2$ is determined by (1.76) for each value of $D \geq 1$. (Another characterization of BES$^{(D)}$ for fractional dimensions D is given by Lamperti's relation (1.130) in Exercise 1.16.)

For BES$^{(D)}$ starting from $x > 0$, denote its first visiting time at the origin by

$$T^x = \inf\{t > 0 : R^x(t) = 0\}. \tag{1.82}$$

The answer to the above question is given by the following theorem.

Theorem 1.1 (i) $D \geq 2 \implies T^x = \infty, {}^\forall x > 0$, *with probability 1.*

(ii) $D > 2 \implies \lim_{t\to\infty} R^x(t) = \infty, {}^\forall x > 0$, *with probability 1, i.e. the process is transient.*

(iii) $D = 2 \implies \inf_{t>0} R^x(t) = 0, {}^\forall x > 0$, *with probability 1.*
That is, BES$^{(2)}$ starting from $x > 0$ does not visit the origin, but it can visit any neighborhood of the origin.

(iv) $1 \leq D < 2 \implies T^x < \infty, {}^\forall x > 0$, *with probability 1, i.e. the process is recurrent.*

Proof For $0 < x_1 < x < x_2 < \infty$, let

$$\sigma = \inf\{t > 0 : R^x(t) = x_1 \text{ or } R^x(t) = x_2\},$$

and define

$$\phi(x) = \phi(x; x_1, x_2) = \mathrm{P}[R^x(\sigma) = x_2].$$

By definition,

$$\phi(x_1) = 0, \quad \phi(x_2) = 1. \tag{1.83}$$

Consider a process

$$M(t) = \phi(R^x(t \wedge \sigma)).$$

It is also written as $M(t) = \mathrm{E}[\phi(R^x(\sigma))|\mathscr{F}_t]$. By the definition of filtration,

$$\begin{aligned}\mathrm{E}[M(t)|\mathscr{F}_s] &= \mathrm{E}[\mathrm{E}[\phi(R^x(\sigma))|\mathscr{F}_t]|\mathscr{F}_s] \\ &= \mathrm{E}[\phi(R^x(\sigma))|\mathscr{F}_s] = M(s), \quad 0 \le^{\forall} s \le t,\end{aligned} \tag{1.84}$$

that is, $M(t)$ is a martingale. Provided that $\phi(x) \in \mathrm{C}_b^2$, we apply Itô's formula using SDE (1.60) of $\mathrm{BES}^{(D)}$ and obtain

$$\begin{aligned}M(t) = \phi(x) &+ \int_0^{t\wedge\sigma} \phi'(R^x(s)) \left[dB(s) + \frac{D-1}{2}\frac{ds}{R^x(s)}\right] \\ &+ \int_0^{t\wedge\sigma} \frac{1}{2}\phi''(R^x(s))\langle dB, dB\rangle_s \\ = \phi(x) &+ \int_0^{t\wedge\sigma} \phi'(R^x(s)) dB(s) + \int_0^{t\wedge\sigma} \frac{1}{2}\left[\phi''(R^x(s)) + \frac{D-1}{R^x(s)}\phi'(R^x(s))\right] ds.\end{aligned}$$

Since $M(t)$ is a martingale, the drift term should be zero. Hence we obtain the following differential equation:

$$\phi''(x) + \frac{D-1}{x}\phi'(x) = 0, \quad x_1 < x < x_2. \tag{1.85}$$

It is equal to $\left(\frac{d}{dx} + \frac{D-1}{x}\right)\phi'(x) = 0$, then with a constant c it is integrated as $\phi'(x) = cx^{-(D-1)}$. With the first of the boundary conditions, $\phi(x_1) = 0$ in (1.83), it is again integrated as

$$\phi(x) = \begin{cases} c\displaystyle\int_{x_1}^{x} y^{-(D-1)} dy = \frac{c}{2-D}(x^{2-D} - x_1^{2-D}), & \text{if } D \neq 2, \\ c\displaystyle\int_{x_1}^{x} \frac{dy}{y} = c(\log x - \log x_1), & \text{if } D = 2. \end{cases}$$

By imposing the second boundary condition $\phi(x_2) = 1$ of (1.83), the integral constant c is determined and we have

$$\phi(x) = \phi(x; x_1, x_2) = \begin{cases} \dfrac{x^{2-D} - x_1^{2-D}}{x_2^{2-D} - x_1^{2-D}}, & \text{if } D \neq 2, \\ \dfrac{\log x - \log x_1}{\log x_2 - \log x_1}, & \text{if } D = 2. \end{cases} \tag{1.86}$$

(i) If $D > 2$, then $2 - D < 0$ and for any $x_2 = L > x$ the upper equation in (1.86) gives

$$\phi(x; 0, L) \equiv \lim_{x_1 \to 0} \phi(x; x_1, L)$$
$$= \lim_{x_1 \to 0} \frac{x^{2-D} - x_1^{2-D}}{L^{2-D} - x_1^{2-D}} = 1.$$

It implies that $\mathrm{BES}^{(D)}$ starting from $x > 0$ will arrive at any positive point $L > 0$ before arriving at the origin with probability 1. Then $T^x = \infty$. For $D = 2$, the lower equation in (1.86) gives

$$\phi(x; 0, L) = \lim_{x_1 \to 0} \frac{\log x - \log x_1}{\log L - \log x_1} = 1.$$

Then we also see that $T^x = \infty$.
(ii) Let $x_k = \alpha^k x$, $k \in \mathbb{N}$ with $\alpha > 1$. When $D > 2$, put $\beta = 2 - D < 0$. The upper equation in (1.86) gives

$$\phi(x_k; x_{k-1}, x_{k+1}) = \frac{x_k^\beta - x_{k-1}^\beta}{x_{k+1}^\beta - x_{k-1}^\beta} = \frac{\alpha^{k\beta} - \alpha^{(k-1)\beta}}{\alpha^{(k+1)\beta} - \alpha^{(k-1)\beta}}$$
$$= \frac{\alpha^\beta - 1}{\alpha^{2\beta} - 1} = \frac{1}{\alpha^\beta + 1} > \frac{1}{2}.$$

Now we consider an asymmetric simple random walk on $\mathbb{Z}$ starting from a site $n > 0$ such that in each step the probability to go right is given by $p = 1/(\alpha^\beta + 1) > 1/2$ and the probability to go left is $1 - p < 1/2$. Since such an asymmetric random walk is transient, by comparing it we can conclude that $R^x(t) \to \infty$ as $t \to \infty$, $^\forall x > 0$ with probability 1.
(iii) In the lower equation for $D = 2$ in (1.86), put $x_1 = 1/n$ and $x_2 = e^n$, $n \in \mathbb{N}$. Then for $x_1 < x < x_2$,

$$\phi(x; 1/n, e^n) = \frac{\log x + \log n}{n + \log n} \longrightarrow 0 \quad \text{as } n \to \infty.$$

It means that for any $n \in \mathbb{N}$, $R^x(t)$, $t \geq 0$ can approach $1/n$ and the statement is concluded.
(iv) If $1 \leq D < 2$, then $\lim_{x_1 \to 0} x_1^{2-D} = 0$. Then the upper equation in (1.86) gives

$$\phi(x; 0, L) = \frac{x^{2-d}}{L^{2-d}} \longrightarrow 0 \quad \text{as } L \to \infty.$$

It implies that $T^x < \infty$ with probability 1. □

1.13 Bessel Flow and Another Critical Dimension $\overline{D}_c = 3/2$

In the previous subsection we defined the $\text{BES}^{(D)}$ for continuous values of dimension $D \geq 1$ and studied the dependence of the probability law of this process on D. Theorem 1.1 states that we have a *critical dimension*,

$$D_c = 2,$$

for competition between the two effects acting on the Bessel process, the 'random force' (the martingale term) and the 'entropy force' (the outward drift term) in (1.60); when $D > D_c$, the latter dominates the former and the process becomes transient, and when $D < D_c$, the former is relevant and recurrence to the origin of the process is realized frequently.

Here we show that there is another critical dimension [6],

$$\overline{D}_c = \frac{3}{2}.$$

In order to characterize the transition at $\overline{D}_c$, we have to investigate the dependence of the behavior of $R^x(t)$ on its initial value, $x > 0$. We call the one-parameter family $\{R^x(t) : t \geq 0\}_{x>0}$ the *Bessel flow* for each fixed $D > 0$.

For $0 < x < y$, we trace the motions of two $\text{BES}^{(D)}$'s starting from x and y by solving (1.60) using the *common* BM, $B(t)$, $t \geq 0$,

$$\begin{aligned} R^x(t) &= x + B(t) + \frac{D-1}{2}\int_0^t \frac{ds}{R^x(s)}, \\ R^y(t) &= y + B(t) + \frac{D-1}{2}\int_0^t \frac{ds}{R^y(s)}, \quad 0 \leq t < T^x. \end{aligned} \tag{1.87}$$

We will see that

$$\begin{aligned} x < y \quad &\Longrightarrow \quad R^x(t) < R^y(t), \quad 0 \leq t < T^x \quad \text{with probability 1} \\ &\Longrightarrow \quad T^x \leq T^y \quad \text{with probability 1}. \end{aligned}$$

The interesting fact is that in the intermediate fractional dimensions, $\overline{D}_c < D < D_c$, it is possible to have a situation where $T^x = T^y$ even for $x < y$. See Fig. 1.4. The main theorem in this section is the following [6].

Theorem 1.2 *For* $0 < x < y < \infty$,

(i) $1 \leq D \leq 3/2 \quad \Longrightarrow \quad T^x < T^y$ *with probability 1.*
(ii) $3/2 < D < 2 \quad \Longrightarrow \quad \text{P}[T^x = T^y] > 0.$

In order to prove the theorem, we first verify the following lemma.

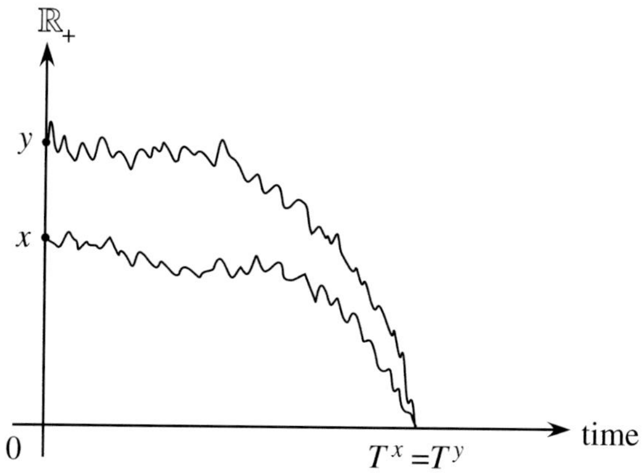

Fig. 1.4 In the intermediate fractional dimensions, $3/2 < D < 2$, there is a positive probability that two Bessel processes starting from different initial positions, $0 < x < y < \infty$, return to the origin simultaneously, $T^x = T^y$

Lemma 1.2 *Let* $0 < x < y$. *The event* $\{\omega : T^x = T^y\}$ *is the same as the event*

$$\left\{\omega : \sup_{0 \le t < T^x} \frac{R^y(t) - R^x(t)}{R^x(t)} < \infty\right\}. \tag{1.88}$$

up to an event of probability 0.

Proof For $x \le y$ put

$$q(x, y) = \mathrm{P}[T^x = T^y].$$

By the scaling property (1.62) of $\mathrm{BES}^{(D)}$, we see that

$$q(x, y) = q(1, y/x).$$

Since $\lim_{r \to \infty} \mathrm{P}[T^r < t] = 0$ for any $t > 0$,

$$\lim_{r \to \infty} q(1, r) = 0. \tag{1.89}$$

Note that

$$\begin{aligned} \text{event (1.88)} \quad &\Longleftrightarrow \quad \left\{\omega : \frac{R^y(t) - R^x(t)}{R^x(t)} \le {}^{\exists}c < \infty, \quad 0 \le t < T^x\right\} \\ &\Longleftrightarrow \quad \left\{\omega : R^y(t) \le (1 + {}^{\exists}c) R^x(t), \quad 0 \le t < T^x\right\}. \end{aligned}$$

Then $R^x(t) = 0 \Longrightarrow R^y(t) = 0$; that is, if (1.88) holds, then $T^x = T^y$. Hence, in order to prove the lemma, it is sufficient to show that the event where $T^x = T^y$ but (1.88) does not hold has probability 0. For $r > 0$ let

$$p_r = \mathrm{P}\left[T^x = T^y \text{ and } \sup_{0 \le t < T^x} \frac{R^y(t) - R^x(t)}{R^x(t)} \ge r\right].$$

Assume that the time $\tau_r = \inf\{0 < t < T^x : (R^y(t) - R^x(t))/R^x(t) = r\} < \infty$, at which $R^y(t)/R^x(t) = 1+r$. We consider the BES$^{(D)}$ started at this time. The strong Markov property of the Bessel process implies that the probability to have $T^x = T^y$ is given by $q(1, 1+r)$ for this restarted Bessel process. The probability p_r is defined for the event that $T^x = T^y$ and $\tau_r < \infty$, then

$$p_r \le q(1, 1+r).$$

On the other hand, (1.89) states $\lim_{r\to\infty} q(1, 1+r) = 0$. Then

$$p_\infty = \lim_{r\to\infty} p_r = \mathrm{P}\left[T^x = T^y \text{ and } \sup_{0\le t<T^x} \frac{R^y(t) - R^x(t)}{R^x(t)} = \infty\right] = 0.$$

The proof is completed. □

Now we prove the theorem.

Proof of Theorem 1.2. For $0 < x < y$, consider a process

$$Z(t) = \log\left(\frac{R^y(t) - R^x(t)}{R^x(t)}\right), \quad t < T^x, \tag{1.90}$$

where $R^x(t)$ and $R^y(t)$ belong to the same Bessel flow and satisfy (1.87). Application of Itô's formula (1.52) leads to (Exercise 1.17)

$$dZ(t) = -\frac{dB(t)}{R^x(t)} + \left[\left(\frac{3}{2} - D\right) + \frac{D-1}{2}\frac{R^y(t) - R^x(t)}{R^y(t)}\right]\frac{dt}{R^x(t)^2}. \tag{1.91}$$

Now we perform the random time change $t \mapsto \tilde{t}$ by

$$\tilde{t} = \int_0^t \frac{ds}{R^x(s)^2} \iff d\tilde{t} = \frac{dt}{R^x(t)^2}. \tag{1.92}$$

For $1 \le D < 2$, we can prove that (Exercise 1.18)

$$\int_0^{T^x} \frac{ds}{R^x(s)^2} = \infty \quad \text{with probability 1}. \tag{1.93}$$

It implies that T^x is mapped to ∞ by this time change. We put

$$\widetilde{B}(\tilde{t}) = -\int_0^t \frac{dB(s)}{R^x(s)},$$

which gives

$$d\langle\widetilde{B}, \widetilde{B}\rangle_{\tilde{t}} = \frac{d\langle B, B\rangle_t}{R^x(t)^2} = \frac{dt}{R^x(t)^2} = d\tilde{t},$$

and thus $\widetilde{B}(\tilde{t})$ is a BM. So we write $\widetilde{Z}(\tilde{t}) = Z(t)$, $\widetilde{R}(\tilde{t}) = R(t)$, and obtain the SDE as

$$d\widetilde{Z}(\tilde{t}) = d\widetilde{B}(\tilde{t}) + \left[\left(\frac{3}{2} - D\right) + \frac{D-1}{2}\frac{\widetilde{R}^y(\tilde{t}) - \widetilde{R}^x(\tilde{t})}{\widetilde{R}^y(\tilde{t})}\right] d\tilde{t}. \tag{1.94}$$

(i) Assume that $1 \le D \le 3/2$. In this case $3/2 - D \ge 0$ and

$$\frac{\widetilde{R}^y(\tilde{t}) - \widetilde{R}^x(\tilde{t})}{\widetilde{R}^y(\tilde{t})} > 0, \quad 0 \le \tilde{t} < \infty.$$

Then the coefficient of drift term in (1.94) is positive, and thus

$$\sup_{0\le\tilde{t}<\infty} \widetilde{Z}(\tilde{t}) = \infty \iff \sup_{0\le t<T^x} e^{Z(t)} = \sup_{0\le t<T^x} \frac{R^y(t) - R^x(t)}{R^x(t)} = \infty.$$

By Lemma 1.2, $\mathrm{P}[T^x = T^y] = 0$ is concluded.
(ii) Assume $3/2 < D < 2$. Choose $D' \in (3/2, D)$ and put $\varepsilon = 2(D - D')/(D - 1)$. Consider the case where $y = (1 + \varepsilon/2)x$ and let

$$\tilde{\sigma} = \inf\left\{\tilde{t} > 0 : \widetilde{R}^y(\tilde{t}) - \widetilde{R}^x(\tilde{t}) = \varepsilon \widetilde{R}^y(\tilde{t})\right\}.$$

Then, for $0 \le \tilde{t} < \tilde{\sigma}$, $(\widetilde{R}^y(\tilde{t}) - \widetilde{R}^x(\tilde{t}))/\widetilde{R}^y(\tilde{t}) \le \varepsilon$, and hence the coefficient of drift term in (1.94) is bounded from above as

$$\left(\frac{3}{2} - D\right) + \frac{D-1}{2}\frac{\widetilde{R}^y(\tilde{t}) - \widetilde{R}^x(\tilde{t})}{\widetilde{R}^y(\tilde{t})} \le \left(\frac{3}{2} - D\right) + \frac{D-1}{2} \times \frac{2(D-D')}{D-1} = \frac{3}{2} - D'.$$

Consider the stochastic process $\widetilde{Z}^*(\tilde{t})$ solving the SDE

$$d\widetilde{Z}^*(\tilde{t}) = d\widetilde{B}(\tilde{t}) + \left(\frac{3}{2} - D'\right) d\tilde{t}, \quad \widetilde{Z}^*(0) = \widetilde{Z}(0) = \log\frac{\varepsilon}{2}. \tag{1.95}$$

Then

$$\widetilde{Z}(\tilde{t}) \le \widetilde{Z}^*(\tilde{t}), \quad 0 \le \tilde{t} < \tilde{\sigma}.$$

Since we have assumed $D' > 3/2$, the coefficient of drift term in (1.95) is negative, and there is a positive probability that $\widetilde{Z}^*(\tilde{t})$ started at $\log(\varepsilon/2)$ never reaches $\log\varepsilon$. On this event, $\widetilde{Z}(\tilde{t}) < \log\varepsilon$. The process $\widetilde{Z}$ is just a time change of the process Z, and so we can conclude that with a positive probability

$$\log\left(\frac{R^y(t) - R^x(t)}{R^x(t)}\right) < \log\varepsilon \iff \frac{R^y(t) - R^x(t)}{R^x(t)} < \varepsilon.$$

The event (1.88) then holds and by Lemma 1.2

$$\mathrm{P}[T^x = T^y] = q\left(x, \left(1+\frac{\varepsilon}{2}\right)x\right) = q\left(1, 1+\frac{\varepsilon}{2}\right) > 0.$$

The proof is completed. □

1.14 Hypergeometric Functions Representing Bessel Flow

In Theorem 1.2 (ii) we proved $\mathrm{P}[T^y = T^x] > 0$ for $0 < x < y < \infty$, when $3/2 < D < 2$. A striking fact is that this probability can be explicitly expressed using *Gauss's hypergeometric function* [6]. Here Gauss's hypergeometric function is defined by

$$F(\alpha, \beta, \gamma; u) = \sum_{n=0}^{\infty} \frac{(\alpha)_n (\beta)_n}{(\gamma)_n} \frac{u^n}{n!},$$

where the Pochhammer symbol, $(c)_0 = 1$, $(c)_n = c(c+1)\cdots(c+n-1), n \geq 1$, is used (see, for instance, Chap. 2 of [1]). It is a fundamental solution at $u = 0$ of *Gauss's hypergeometric equation*

$$u(1-u)F'' + \Big\{\gamma - (\alpha+\beta+1)u\Big\}F' - \alpha\beta F = 0. \tag{1.96}$$

The following is known as *Gauss's summation formula*,

$$F(\alpha, \beta, \gamma; 1) = \frac{\Gamma(\gamma)\Gamma(\gamma-\alpha-\beta)}{\Gamma(\gamma-\alpha)\Gamma(\gamma-\beta)}, \quad \Re\gamma > 0, \Re(\Gamma - \alpha - \beta) > 0. \tag{1.97}$$

Proposition 1.1 *For* $3/2 < D < 2, 0 < x < y < \infty$,

$$\mathrm{P}[T^x = T^y] = 1 - \mathrm{P}[T^x < T^y], \tag{1.98}$$

with

$$\mathrm{P}[T^x < T^y] = \frac{\Gamma(D-1)}{\Gamma(2D-3)\Gamma(2-D)} \int_0^{(y-x)/y} \frac{du}{(1-u)^{D-1} u^{2(2-D)}} \tag{1.99}$$

$$= \frac{\Gamma(D-1)}{\Gamma(2(D-1))\Gamma(2-D)} \left(\frac{y-x}{y}\right)^{2D-3}$$

$$\times F\left(2D-3, D-1, 2(D-1); \frac{y-x}{y}\right). \tag{1.100}$$

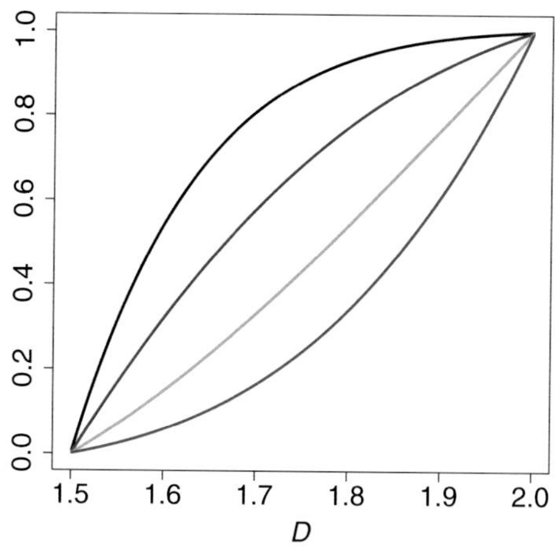

Fig. 1.5 Probabilities $\mathrm{P}[T^x = T^y]$ are plotted as functions of the dimension D for $x \equiv 1$ and $y = 1.1$ (*black*), 2.0 (*blue*), 10 (*green*), and 100 (*red*), respectively

Numerical values of $\mathrm{P}[T^x = T^y]$ are plotted in Fig. 1.5 as functions of the dimension D for several values of x and y.

Proof of Proposition 1.1. By the scaling property of Bessel process (1.62), we can replace the variables as $x \to 1$ and $y \to 1 + x, x > 0$ without loss of generality. For $x > 0$ consider a process

$$S(t) = \frac{R^{1+x}(t) - R^1(t)}{R^1(t)}, \quad t \geq 0. \tag{1.101}$$

Itô's formula (1.52) gives (Exercise 1.19)

$$dS(t) = -\frac{S(t)}{R^1(t)} dB(t) + \left[\frac{3-D}{2} \frac{1}{S(t)} - \frac{D-1}{2} \frac{1}{S(t)(1+S(t))} \right] \left(\frac{S(t)}{R^1(t)} \right)^2 dt. \tag{1.102}$$

Here we perform the random time change $t \to \bar{t}$ by

$$\bar{t} = \int_0^t \left(\frac{S(s)}{R^1(s)} \right)^2 ds. \tag{1.103}$$

If we set

$$\overline{B}(\bar{t}) = -\int_0^t \frac{S(s)}{R^1(s)} dB(s), \tag{1.104}$$

it is a BM. Hence, if we set $\overline{S}(\bar{t}) = S(t)$, then we obtain the following SDE,

$$\begin{aligned} d\overline{S}(\bar{t}) &= d\overline{B}(\bar{t}) + \left[\frac{3-D}{2} \frac{1}{\overline{S}(\bar{t})} - \frac{D-1}{2} \frac{1}{\overline{S}(\bar{t})(\overline{S}(\bar{t})+1)} \right] d\bar{t} \\ &= d\overline{B}(\bar{t}) + \left[\frac{2-D}{\overline{S}(\bar{t})} + \frac{D-1}{2} \frac{1}{\overline{S}(\bar{t})+1} \right] d\bar{t}. \end{aligned} \tag{1.105}$$

Let

$$\psi(x) = \mathrm{P}[T^1 = T^{1+x}] = q(1, 1+x), \tag{1.106}$$

and consider the process

$$\overline{M}(\bar{t}) = \psi(\overline{S}(\bar{t})) = \mathrm{P}\left[T^1 = T^{\overline{R}^{1+x}(\bar{t})/\overline{R}^1(\bar{t})}\right], \tag{1.107}$$

where $\overline{R}^x(\bar{t}) = R^x(t), t \geq 0$. By Itô's formula (1.52) with (1.105), we obtain the SDE for (1.107) as

$$d\overline{M}(\bar{t}) = \psi'(\overline{S}(\bar{t}))d\overline{B}(\bar{t}) + \psi'(\overline{S}(\bar{t}))\left[\frac{2-D}{\overline{S}(\bar{t})} + \frac{D-1}{2}\frac{1}{\overline{S}(\bar{t})+1}\right]d\bar{t} + \frac{1}{2}\psi''(\overline{S}(\bar{t}))d\bar{t}. \tag{1.108}$$

On the other hand, the scaling property of the Bessel process (1.62) gives

$$\left(R^{\overline{R}^{1+x}(\bar{t})/\overline{R}^1(\bar{t})}(t)\right)_{t\geq 0} \overset{\text{(law)}}{=} \left(\frac{1}{c}R^{1+x}(c^2 t)\right)_{t\geq 0} \quad \text{with} \quad c = \frac{1+x}{\overline{R}^{1+x}(\bar{t})/\overline{R}^1(\bar{t})}$$

for any $\bar{t} \geq 0$, and thus $\overline{M}(\bar{t})$ is a martingale

$$\mathrm{E}[\overline{M}(\bar{t})|\mathscr{F}_{\bar{s}}] = \overline{M}(\bar{s}), \quad 0 \leq^{\forall} \bar{s} < \bar{t}. \tag{1.109}$$

Then the drift term in (1.108) should be zero and a differential equation for $\psi(x)$ is obtained:

$$\frac{1}{2}\psi''(x) + \left[\frac{2-D}{x} + \frac{D-1}{2}\frac{1}{x+1}\right]\psi'(x) = 0. \tag{1.110}$$

If we set $\varphi(x) = \psi'(x)$, (1.110) becomes

$$\varphi'(x) = -\left[2(2-D)\frac{1}{x} + (D-1)\frac{1}{1+x}\right]\varphi(x),$$

which is solved by $\varphi(x) \equiv 0$ and

$$\varphi(x) = \frac{1}{x^{2(2-D)}(1+x)^{D-1}}.$$

Hence, we obtain the two solutions of (1.110): $\psi(x) = 1$ and

$$\psi(x) = \int_0^x \frac{dy}{y^{2(2-D)}(1+y)^{D-1}} = \int_0^{x/(1+x)} \frac{du}{(1-u)^{D-1}u^{2(2-D)}},$$

where we have assumed $\psi(0) = 0$ for the latter solution and set $y = u/(1-u)$ in the integral. Then the general solution of (1.110) is given by

$$\psi(x) = c_1 + c_2 \int_0^{x/(1+x)} \frac{du}{(1-u)^{D-1}u^{2(2-D)}}.$$

By the conditions

$$\psi(0) = \mathrm{P}[T^1 = T^1] = 1,$$
$$\psi(\infty) = \lim_{x\to\infty} \mathrm{P}[T^{1+x} = T^1] = 0,$$

we obtain

$$\psi(x) = 1 - \frac{\Gamma(D-1)}{\Gamma(2D-3)\Gamma(2-D)} \int_0^{x/(1+x)} \frac{du}{(1-u)^{D-1}u^{2(2-D)}},$$

where we have used the integral formula

$$\int_0^1 u^{p-1}(1-u)^{q-1}du = \frac{\Gamma(p)\Gamma(q)}{\Gamma(p+q)}, \quad \Re p > 0, \quad \Re q > 0. \tag{1.111}$$

By putting $1/(1+x) \to x/y$, (1.99) is obtained. For (1.100), we change the variable in (1.110) as $x \to u$ with

$$x = \frac{u}{1-u} \iff u = \frac{x}{1+x},$$

and put $\widetilde{\psi}(u) = \psi(x)$. Then we have the equation

$$u(1-u)\widetilde{\psi}''(u) + \Big\{2(2-D) - (3-D)u\Big\}\widetilde{\psi}'(u) = 0. \tag{1.112}$$

It is a special case of Gauss's hypergeometric equation (1.96) with setting parameters

$$\alpha = 0, \quad \beta = 2-D, \quad \gamma = 2(2-D). \tag{1.113}$$

For Gauss's hypergeometric equation, we can adopt $\{F(\alpha, \beta, \gamma; u), u^{1-\gamma}F(1-\gamma+\alpha, 1-\gamma+\beta, 2-\gamma; u)\}$ as a fundamental system of solutions at $u = 0$. Under (1.113), the former is 1 and the latter is $u^{2D-3}F(2D-3, D-1, 2(D-1); u)$, and hence the solution will be expressed by

$$\widetilde{\psi}(u) = \widetilde{c}_1 + \widetilde{c}_2 u^{2D-3} F(2D-3, D-1, 2(D-1); u)$$

with constants $\widetilde{c}_1$ and $\widetilde{c}_2$. By the conditions

$$\widetilde{\psi}(0) = \psi(0) = \mathrm{P}[T^1 = T^1] = 1,$$
$$\widetilde{\psi}(1) = \psi(\infty) = \lim_{x\to\infty} \mathrm{P}[T^{1+x} = T^1] = 0, \tag{1.114}$$

they are determined as

$$\widetilde{c}_1 = 1$$
$$\widetilde{c}_2 = -F(2D-3, D-1, 2(D-1); 1)^{-1} = -\frac{\Gamma(D-1)}{\Gamma(2(D-1))\Gamma(2-D)},$$

where (1.97) is used. We change the variable $u \to x = u/(1-u)$, and then replace $1/(1+x)$ by x/y. The formula (1.100) is thus obtained and the proof is completed. □

Exercises

1.1 Using the transition probability density function (1.1), the LHS of (1.10) is written as $\int_{-\infty}^{\infty}(y^2-t)p(t-s, y|x)dy$, under the condition $B(s) = x$. Show that it is equal to $x^2 - s$ and directly prove (1.10).

1.2 Consider the integrals $I_{nm} = \int_{-\infty}^{\infty} e^{-x^2} H_n(x)H_m(x)dx$, $n, m \in \mathbb{N}_0$.
(i) By (1.19), they are written as

$$I_{nm} = (-1)^n \int_{-\infty}^{\infty} \frac{d^n e^{-x^2}}{dx^n} H_m(x)dx, \quad n, m \in \mathbb{N}_0. \tag{1.115}$$

Suppose $n > m$. Then prove that $I_{nm} = 0$.
(ii) Show that $I_{nn} = 2^n n!\sqrt{\pi}$.
The above proves that $\{H_n(x)\}_{n\in\mathbb{N}_0}$ have the orthogonality property

$$\int_{-\infty}^{\infty} e^{-x^2} H_n(x)H_m(x)dx = 2^n n!\sqrt{\pi}\delta_{nm}, \quad n, m \in \mathbb{N}_0. \tag{1.116}$$

1.3 Derive the formula

$$\sum_{n=0}^{\infty} H_n(z)\frac{s^n}{n!} = e^{2sz-s^2} \tag{1.117}$$

from (1.17) with (1.15) and (1.18). Show the following contour integral representations of the Hermite polynomials,

$$H_n(z) = \frac{n!}{2\pi\sqrt{-1}} \oint_{C(\delta_0)} d\eta \, \frac{e^{2\eta z-\eta^2}}{\eta^{n+1}}, \quad n \in \mathbb{N}_0, \tag{1.118}$$

where $C(\delta_0)$ is a closed contour on the complex plane $\mathbb{C}$ encircling the origin 0 once in the positive direction.

1.4 Let $F(s,z) = e^{2sz-s^2}$.
(i) Show that $F(s,z)$ satisfies

$$\frac{\partial F}{\partial s} - (2z - 2s)F = 0, \tag{1.119}$$
$$\frac{\partial F}{\partial z} - 2sF = 0. \tag{1.120}$$

(ii) Using the formula (1.117), derive the following *recurrence relations*:

$$H_{n+1}(z) - 2zH_n(z) + 2nH_{n-1}(z) = 0, \tag{1.121}$$
$$H'_n(z) = 2nH_{n-1}(z), \quad n \in \mathbb{N}, \tag{1.122}$$

where $H'_n(z) = dH_n(z)/dz$.
(iii) From (1.121) and (1.122), derive the equations

$$H''_n(z) - 2zH'_n(z) + 2nH_n(z) = 0, \quad n \in \mathbb{N}_0.$$

That is, $\{H_n(z)\}_{n \in \mathbb{N}_0}$ satisfy the differential equation

$$u'' - 2zu' + 2nu = 0. \tag{1.123}$$

This is known as the *Hermite differential equation.*

1.5 Show

$$\mathrm{E}[(B(t) - B(s))^4] = 3(t-s)^2, \quad 0 \le s \le t. \tag{1.124}$$

1.6 For any random variable X with $|\mathrm{E}[X]| < \infty$, prove Chebyshev's inequality,

$$\mathrm{P}[|X - \mathrm{E}[X]| > \varepsilon] \le \frac{1}{\varepsilon^2}\mathrm{E}[|X - \mathrm{E}[X]|^2], \quad {}^\forall\varepsilon > 0.$$

1.7 Let $B_i(t),\ t \ge 0, i = 1, 2, \dots, D$ be independent BMs. Prove that $\langle B_i, B_j \rangle_t = \delta_{ij}t,\ 1 \le i, j \le D, t \ge 0$.

1.8 For $n \in \mathbb{N}_0$, let $\widehat{m}_n(t,x) = (t/2)^{n/2}u_n(x/\sqrt{2t})$, $(t,x) \in [0,\infty) \times \mathbb{R}$. Assume that $u_n(z) \in \mathrm{C}^2_{\mathrm{b}}$. Let $B(t),\ t \ge 0$ be BM. By applying Itô's formula (1.52), show that $\widehat{m}_n(t, B(t)),\ t \ge 0$ is a local martingale, if and only if $u_n(z)$ satisfies the Hermite differential equation (1.123).

1.9 Prove the scaling property (1.62) for $\mathrm{BES}^{(D)}$.

1.10 For $x > 0$, let $\tau_1^x = \inf\{t > 0 : R^x(t) = x/2\}$ and $\tau_2^x = \inf\{t > 0 : R^x(t) = x/2^2\}$. Define

$$I_1 = \int_0^{\tau_1^x} \frac{dt}{R^x(t)^2}, \quad I_2 = \int_{\tau_1^x}^{\tau_2^x} \frac{dt}{R^x(t)^2}.$$

Prove that I_1 and I_2 are *independently and identically distributed (i.i.d.)*.

1.11 Confirm that $p^{(D)}(t, y|x)$ given by (1.75) satisfies the backward Kolmogorov equation (1.71) with (1.70) for $\text{BES}^{(D)}$.

1.12 Show that, for $x \in \mathbb{R}_+, t > 0$, $p^{(3)}(t, y|x)$ given by (1.77) is well-normalized as $\int_{\mathbb{R}_+} p^{(3)}(t, y|x)dy = 1$.

1.13 Take the limit $x \to 0$ in (1.77) and obtain the formula

$$\lim_{x\to 0} p^{(3)}(t, y|x) = \frac{2}{t} y^2 p(t, y|0) \equiv p^{(3)}(t, y|0). \tag{1.125}$$

It coincides with the result obtained from the middle formula in (1.75) by putting $D = 3 \Leftrightarrow \nu = 1/2$, since $\Gamma(3/2) = \sqrt{\pi}/2$.

1.14 Consider an absorbing Brownian motion started at $x > 0$ with an absorbing wall at the origin. Let τ be the time the BM is absorbed: $\tau = \inf\{t > 0 : B^x(t) = 0\}$. For given $T > 0$, the probability that the BM is not yet absorbed and thus survives at that time T is given by $\mathrm{P}^x[\tau > T] = \int_0^\infty q^{\text{abs}}(T, y|x)dy$, which will be called the *survival probability* up to time T.
(i) Prove the long-term asymptotics of the survival probability,

$$\mathrm{P}^x[\tau > T] \sim \sqrt{\frac{2}{\pi}} x T^{-1/2} \quad \text{as } T \to \infty. \tag{1.126}$$

(ii) Consider the absorbing Brownian motion under the condition that it survives up to a given time $T > 0$. In other words, it is the one-dimensional Brownian motion conditioned to stay positive up to time T. For $0 < t < T$, the transition probability density of such conditional process started at $x > 0$ at time 0 and arriving at $y > 0$ at time t will be given by

$$p_T^{\text{positive}}(t, y|x) = \frac{q^{\text{abs}}(t, y|x)\mathrm{P}^y[\tau > T - t]}{\mathrm{P}^x[\tau > T]}. \tag{1.127}$$

Take the limit $T \to \infty$ of this transition probability density function.

1.15 The RHS of (1.34) is a special case of the *Girsanov transformation* [7],

$$\mathrm{E}\left[F(B(t)) \exp\left\{ \int_s^t \varphi(B(u))dB(u) - \frac{1}{2}\int_s^t \varphi(B(u))^2 du \right\} \middle| \mathscr{F}_s \right] \tag{1.128}$$

with $\varphi(x) \equiv b$. Set $s = 0$, $B(0) = x \in \mathbb{R}_+$ and put

$$\varphi(x) = \frac{1}{x}. \tag{1.129}$$

Show that (1.128) then gives the RHS of (1.80).

1.16 Let $\nu \in \mathbb{R}$ and consider a BM with a constant drift ν, $B^y(t) + \nu t$, which starts from $y \in \mathbb{R}$ at time $t = 0$. The *geometric Brownian motion* with drift ν is defined as $\exp(B^y(t) + \nu t)$, $t \geq 0$. For each $t \geq 0$, we define the random time change $t \mapsto A(t)$ by $A(t) = \int_0^t \exp\{2(B^y(s) + \nu s)\}ds$, and let $R^x(A(t))$ be the $\mathrm{BES}^{(D)}$ with $D = 2(\nu + 1)$ at time $A(t)$ starting from $x = e^y$. Prove that

$$(R^x(A(t)))_{t\geq 0} \stackrel{\text{(law)}}{=} \left(\exp(B^y(t) + \nu t)\right)_{t\geq 0}. \tag{1.130}$$

This formula is called *Lamperti's relation* (see (1.28) in Chap. 11 of [7]).

1.17 Derive (1.91) by applying Itô's formula to (1.90) with (1.87).

1.18 Prove (1.93) for $1 \leq D < 2$ following the instructions below.
(i) For $x > 0$, let $\tau_0^x = 0$, $\tau_n^x = \inf\{t > 0 : R^x(t) = x/2^n\}$, $n \in \mathbb{N}$, and

$$I_n = \int_{\tau_{n-1}^x}^{\tau_n^x} \frac{ds}{R^x(s)^2}, \quad n \in \mathbb{N}.$$

Prove that I_n, $n \in \mathbb{N}$ are i.i.d.
(ii) For $1 \leq D < 2$, conclude (1.93).

1.19 Derive (1.102) by applying Itô's formula to (1.101).

References

1. Andrews, G.E., Askey, R., Roy, R.: Special Functions. Cambridge University Press, Cambridge (1999)
2. Doob, J.L.: Classical Potential Theory and its Probabilistic Counterpart. Springer, Berlin (1984)
3. Karatzas, I., Shreve, S.E.: Brownian Motion and Stochastic Calculus, 2nd edn. Springer, New York (1991)
4. Katori, M.: Determinantal martingales and noncolliding diffusion processes. Stoch. Proc. Appl. **124**, 3724–3768 (2014)
5. Katori, M., Tanemura, H.: Noncolliding processes, matrix-valued processes and determinantal processes. Sugaku Expositions (AMS) **24**, 263–289 (2011)
6. Lawler, G.F.: Conformally Invariant Processes in the Plane. American Mathematical Society, Providence (2005)
7. Revuz, D., Yor, M.: Continuous Martingales and Brownian Motion, 3rd edn. Springer, Berlin (1999)
8. Schoutens, W.: Stochastic Processes and Orthogonal Polynomials. Lecture Notes in Statistics, vol. 146. Springer, New York (2000)

Chapter 2
Schramm–Loewner Evolution (SLE)

Abstract We consider the Loewner chain, which is a time evolution of a conformal transformation defined on the upper-half complex plane. The chain is driven by a given continuous real function of time t and it determines a path γ in the upper half-plane parameterized by t. Schramm–Loewner evolution (SLE) is a stochastic version of the Loewner chain such that the driving function is given by a time change of one-dimensional Brownian motion and thus the path becomes stochastic. We introduce the SLE as a complexification of the Bessel flow studied in Chap. 1. Then the parameter κ of SLE, which is originally introduced to control the time change of Brownian motion driving the SLE, is related to the dimension D of Bessel process. Corresponding to the existence of two critical dimensions $D_c = 2$ and $\overline{D}_c = 3/2$, the appearance of three different phases of the SLE path is clarified. Moreover, based on the detailed analysis of the Bessel flow in $\overline{D}_c < D < D_c$ given in Chap. 1, Cardy's formula for the critical percolation model is derived. We give a list showing correspondence (up to a conjecture) between lattice paths studied in statistical mechanics and SLE paths describing their scaling limits.

2.1 Complexification of Bessel Flow

In the sequel we consider an extension of the Bessel flow $\{R^x(t) : t \geq 0\}_{x>0}$ defined on $\mathbb{R}_+$ to a flow on the upper-half complex plane $\mathbb{H} = \{z = x + \sqrt{-1}y : x \in \mathbb{R},\ y > 0\}$ and its boundary $\partial\mathbb{H} = \mathbb{R}$. Let $\overline{\mathbb{H}} = \mathbb{H} \cup \mathbb{R}$. We set $Z^z(t) = X^z(t) + \sqrt{-1}Y^z(t) \in \overline{\mathbb{H}} \setminus \{0\},\ t \geq 0$ and complexificate (1.60) as

$$dZ^z(t) = dB(t) + \frac{D-1}{2}\frac{dt}{Z^z(t)} \tag{2.1}$$

with the initial condition

$$Z^z(0) = z = x + \sqrt{-1}y \in \overline{\mathbb{H}} \setminus \{0\}.$$

M. Katori, *Bessel Processes, Schramm–Loewner Evolution, and the Dyson Model*, SpringerBriefs in Mathematical Physics 11, DOI 10.1007/978-981-10-0275-5_2

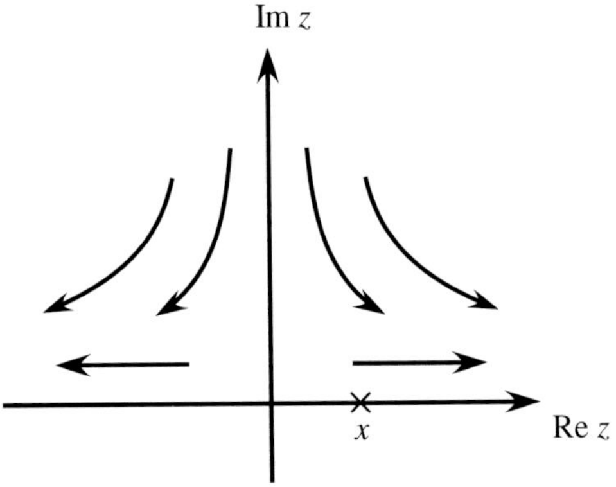

Fig. 2.1 A schematic picture of 'complexificated Bessel flow' on $\overline{\mathbb{H}} \setminus \{0\}$ for $D > 2$

The crucial point of this complexification of Bessel flow is that the BM remains real, $B(t) \in \mathbb{R},\ t \geq 0$. Then, there is an asymmetry between the real part and the imaginary part of the flow in $\mathbb{H}$,

$$dX^z(t) = dB(t) + \frac{D-1}{2}\frac{X^z(t)}{(X^z(t))^2 + (Y^z(t))^2}dt, \tag{2.2}$$

$$dY^z(t) = -\frac{D-1}{2}\frac{Y^z(t)}{(X^z(t))^2 + (Y^z(t))^2}dt. \tag{2.3}$$

Assume $D > 1$. Then as indicated by the minus sign in the RHS of (2.3), the flow is downward in $\overline{\mathbb{H}}$. If the flow goes down and arrives at the real axis, the imaginary part vanishes, $Y^z(t) = 0$, then Eq. (2.2) is reduced to be the same equation as Eq. (1.60) for the $\text{BES}^{(D)}$, which is now considered for $\mathbb{R} \setminus \{0\} = \mathbb{R}_+ \cup \mathbb{R}_-$. If $D > D_c = 2$, by Theorem 1.1 (ii), the flow on $\mathbb{R} \setminus \{0\}$ is asymptotically outward, $X^z(t) \to \pm\infty$ as $t \to \infty$. Therefore, the flow on $\overline{\mathbb{H}}$ will be described as shown by Fig. 2.1. The behavior of flow should be, however, more complicated when $\overline{D}_c = 3/2 < D < D_c$ and $1 < D < \overline{D}_c$.

For $z \in \overline{\mathbb{H}} \setminus \{0\},\ t \geq 0$, let

$$g_t(z) = Z^z(t) - B(t). \tag{2.4}$$

Since $B^x(t)$ and $Z^z(t)$ are stochastic processes, they are considered as functions of time $t \geq 0$, where the initial values x and z are put as superscripts ($B(t) \equiv B^0(t)$). On the other hand, as explained below, g_t is considered as a conformal transformation from a domain $\mathsf{H}_t \subset \mathbb{H}$ to $\mathbb{H}$, and thus it is described as a function of $z \in \mathsf{H}_t$; $g_t(z)$, where time t is a parameter and put as a subscript.

Then, Eq. (2.1) is rewritten for $g_t(z)$ as

$$\frac{\partial g_t(z)}{\partial t} = \frac{D-1}{2}\frac{1}{g_t(z) + B(t)}, \quad t \geq 0 \tag{2.5}$$

with the initial condition

$$g_0(z) = z \in \overline{\mathbb{H}} \setminus \{0\}. \tag{2.6}$$

For each $z \in \overline{\mathbb{H}} \setminus \{0\}$, set

$$\begin{aligned} T^z &= \inf\{t > 0 : Z^z(t) = 0\} \\ &= \inf\{t > 0 : g_t(z) + B(t) = 0\}, \end{aligned} \tag{2.7}$$

and then the solution of Eq. (2.5) exists up to time T^z. For $t \geq 0$ we put

$$\mathsf{H}_t = \{z \in \mathbb{H} : T^z > t\}. \tag{2.8}$$

This ordinary differential equation (2.5) involving the BM is nothing but the celebrated *Schramm–Loewner evolution* (SLE) [13, 17]. It is known that [13], for each $t \geq 0$, the solution $g_t(z)$ of (2.5) gives a unique *conformal transformation* from H_t to $\mathbb{H}$:

$$g_t(z) : \quad \mathsf{H}_t \mapsto \mathbb{H}, \quad \text{conformal},$$

such that

$$g_t(z) = z + \frac{a(t)}{z} + \mathscr{O}\left(\frac{1}{|z|^2}\right), \quad z \to \infty$$

with

$$a(t) = \frac{D-1}{2}t.$$

The usual parameter for the SLE is given by $\kappa > 0$ [13, 17], which is related to D by

$$\kappa = \frac{4}{D-1} \quad \Longleftrightarrow \quad D = 1 + \frac{4}{\kappa}. \tag{2.9}$$

If we set $\widehat{g}_t(z) = \sqrt{\kappa} g_t(z)$ in (2.5), we have the equation in the form [17]

$$\frac{\partial \widehat{g}_t(z)}{\partial t} = \frac{2}{\widehat{g}_t(z) - U_t} \tag{2.10}$$

with

$$U_t = -\sqrt{\kappa} B(t), \quad t \geq 0. \tag{2.11}$$

In the complex analysis, given a real function U_t of $t \geq 0$, a one-parameter family of conformal transformations $(g_t)_{t\geq 0}$ defined by the unique solution of (2.10) under $g_0(z) \equiv z \in \mathbb{H}$ is called the *Loewner chain* driven by $(U_t)_{t\geq 0}$. Note that

$$(-\sqrt{\kappa}B(t))_{t\geq 0} \overset{\text{(law)}}{=} (B(\kappa t))_{t\geq 0}$$

by the left-right symmetry and the scaling property (1.4) of BM. The parameter $\kappa > 0$ is the diffusion constant and it 'speeds up' (as $\kappa \uparrow$) and 'slows down' (as $\kappa \downarrow$) the one-dimensional Brownian motion which drives the stochastic Loewner chain. In the present book, however, we will discuss the SLE using the parameter $D > 1$, since we would like to discuss it as a complexification of the D-dimensional Bessel flow.

The inverse map

$$f_t(z) \equiv g_t^{-1}(t), \quad : \mathbb{H} \mapsto \mathsf{H}_t, \quad t \geq 0, \tag{2.12}$$

is also conformal. The equation of $(f_t(z))_{t\geq 0}$ is then obtained as (Exercise 2.1)

$$\frac{\partial f_t(z)}{\partial t} = -\frac{D-1}{2}\frac{\partial f_t(z)}{\partial z}\frac{1}{z+B(t)}, \quad t \geq 0. \tag{2.13}$$

We call this partial differential equation the *backward SLE*. For (2.10), the inverse map $\widehat{f}_t(z) \equiv \widehat{g}_t^{-1}(z)$ satisfies

$$\frac{\partial \widehat{f}_t(z)}{\partial t} = -\frac{\partial \widehat{f}_t(z)}{\partial z}\frac{2}{z-U_t}, \quad t \geq 0 \tag{2.14}$$

with (2.11).

By the definition of T^z, (2.7), for each $z \in \mathbb{H}$, $Z^z(t) = g_t(z) + B(t) \to 0$ as $t \uparrow T^z$. (In this limit the Eq. (2.5) becomes ill-defined.) Set $\zeta = g_t(z) + B(t)$ provided $t < T^z \iff z \in \mathsf{H}_t$. In this case $g_t(z) \in \mathbb{H}$, $B(t) \in \mathbb{R}$, and hence $\zeta \in \mathbb{H}$. Therefore, an approach $Z^z(t) \to 0$ corresponds to a limit $\zeta \to 0, \zeta \in \mathbb{H}$. Since $\zeta = g_t(z) + B(t) \iff z = g_t^{-1}(\zeta - B(t))$, the behavior of $Z^z(t) \to 0$ will be represented by the limit

$$\gamma(t) \equiv \lim_{\substack{\zeta\to 0,\\ \zeta\in\mathbb{H}}} g_t^{-1}(\zeta - B(t)). \tag{2.15}$$

Using properties of BM and the conformal transformation generalized by the Loewner chain (2.5), Rohde and Schramm [16] proved that $\gamma = \gamma[0, \infty) \equiv \{\gamma(t) : t \in [0, \infty)\} \in \overline{\mathbb{H}}$ is a continuous path with probability 1 running from $\gamma(0) = 0$ to $\gamma(\infty) = \infty$. The path γ obtained from the SLE with the parameter $D > 1$ is called the *SLE*$^{(D)}$ *path*. (See Exercise 2.2.)

2.2 Schwarz–Christoffel Formula and Loewner Chain

In mathematical physics, the *Schwarz–Christoffel formula* may be more popular than the Loewner chain, when conformal transformations are studied. In this section, we discuss the Loewner chain from the viewpoint of the Schwarz–Christoffel formula using a simple example of conformal transformation. We will use the Schwarz–Christoffel transformation in Sect. 2.4, where Cardy's formula in Carleson's form is given for an equilateral triangular domain.

Let Γ be a polygon having vertices $w_1, w_2, \dots, w_n$ and interior angles $\alpha_1\pi, \alpha_2\pi, \dots, \alpha_n\pi$ in the counterclockwise direction and D be the interior of Γ as shown in Fig. 2.2. The following theorem is known as the Schwarz–Christoffel formula [8].

Theorem 2.1 *Let $\widehat{f}$ be any conformal map from $\mathbb{H}$ to D with $\widehat{f}(x_i) = w_i$, $1 \le i \le n-1$, and $\widehat{f}(\infty) = w_n$, where $x_i \in \mathbb{R}$, $1 \le i \le n-1$. Then*

$$\frac{d\widehat{f}(z)}{dz} = C\prod_{i=1}^{n-1}(z - x_i)^{\alpha_i - 1}, \tag{2.16}$$

where C is a complex constant.

As an application of this formula, we consider a conformal map $\widehat{f}$ from $\mathbb{H}$ to the upper-half complex plane with a straight slit starting from the origin: $\mathbb{H} \setminus \{\text{a slit}\}$. Let $0 < \alpha < 1$. As shown by Fig. 2.3, the angle between the slit and the positive

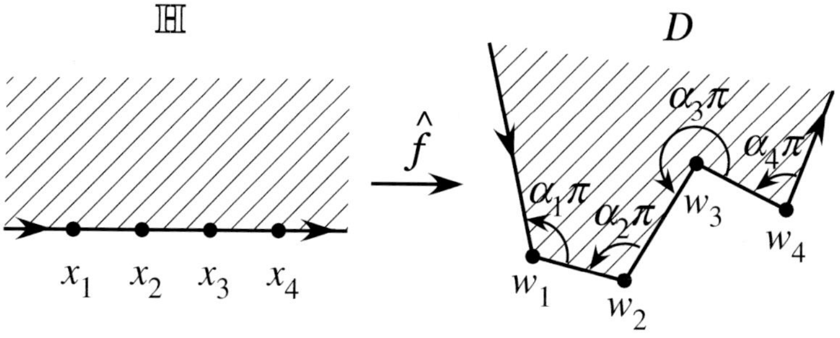

Fig. 2.2 The conformal map $\widehat{f}$ from $\mathbb{H}$ to D with $\widehat{f}(x_i) = w_i$, $1 \le i \le 4$ and $\widehat{f}(\infty) = w_5 \equiv \infty$

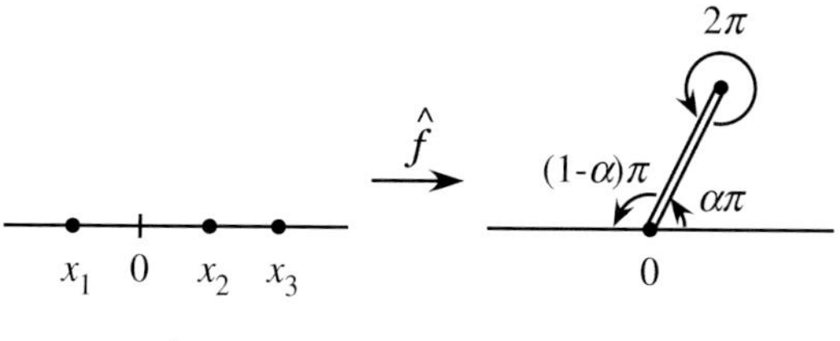

Fig. 2.3 The conformal map $\widehat{f}$ from $\mathbb{H}$ to $\mathbb{H} \setminus \{\text{a slit}\}$, where the angle between the slit and the positive direction of the real axis is $\alpha\pi$, $\alpha \in (0, 1)$

direction of the real axis is supposed to be $\alpha\pi$. Since the region $\mathbb{H} \setminus \{\text{a slit}\}$ can be regarded as a polygon with the interior angles $(1-\alpha)\pi$ on the left side of the origin, 2π around the tip of the slit, and $\alpha\pi$ on the right side of the origin, for any length of a slit, the formula (2.16) gives

$$\frac{d\widehat{f}(z)}{dz} = C(z-x_1)^{-\alpha}(z-x_2)(z-x_3)^{\alpha-1}, \tag{2.17}$$

where $x_1 < 0$, $x_1 < x_2 < x_3$, and $x_3 > 0$. We assume that $\widehat{f}(x_1) = \widehat{f}(x_3) = 0$ and $\widehat{f}(x_2)$ gives the tip of the slit. If we impose the condition on the asymptotics as

$$\frac{\widehat{f}(z)}{z} \to 1 \quad \text{as } z \to \infty,$$

the solution of (2.17) is uniquely determined as

$$\widehat{f}(z) = (z-x_1)^{1-\alpha}(z-x_3)^{\alpha}, \tag{2.18}$$

where the following relation should be satisfied,

$$x_3 - x_2 = \alpha(x_3 - x_1). \tag{2.19}$$

Using (2.18), the Schwarz–Christoffel differential equation (2.17) is rewritten as

$$\frac{d\widehat{f}(z)}{dz}\frac{2}{z-x_2} = \frac{2\widehat{f}(z)}{(z-x_1)(z-x_3)}. \tag{2.20}$$

We then introduce a parameter $t \geq 0$ and assume $x_i = x_i(t)$, $i = 1, 2, 3$, and put $\widehat{f}_t(z) = (z-x_1(t))^{1-\alpha}(z-x_3(t))^{\alpha}$. The differential of $\widehat{f}_t$ with respect to t is given as

$$\frac{\partial \widehat{f}_t(z)}{\partial t} = -\frac{2A_t(z)}{(z-x_1(t))(z-x_3(t))}\widehat{f}_t(z) \tag{2.21}$$

with

$$A_t(z) = \frac{1}{2}\left\{(1-\alpha)(z-x_3(t))\frac{dx_1(t)}{dt} + \alpha(z-x_1(t))\frac{dx_3(t)}{dt}\right\}.$$

Let $x_1(t) = -2ct^{\beta}$, $x_3(t) = (2/c)t^{\beta}$ with constants $c, \beta > 0$. Then we find that, if and only if

$$c = \sqrt{\frac{\alpha}{1-\alpha}}, \quad \beta = \frac{1}{2}, \tag{2.22}$$

$A_t(z)$ becomes independent both of z and t; $A_t(z) \equiv 1$. In this case (2.20) and (2.21) give the equation

$$\frac{\partial \widehat{f}_t(z)}{\partial t} = -\frac{\partial \widehat{f}_t(z)}{\partial z} \frac{2}{z - x_2(t)}, \quad t \geq 0 \tag{2.23}$$

with

$$x_2(t) = \begin{cases} \sqrt{\kappa t}, & \text{if } \alpha \leq 1/2, \\ -\sqrt{\kappa t}, & \text{if } \alpha > 1/2, \end{cases} \tag{2.24}$$

where

$$\kappa = \kappa(\alpha) = \frac{4(1-2\alpha)^2}{\alpha(1-\alpha)}. \tag{2.25}$$

Equation (2.23) can be regarded as the backward Loewner evolution (2.14) driven by (2.24). The obtained conformal transformation

$$\widehat{f}_t(z) = \left(z + 2\sqrt{\frac{\alpha}{1-\alpha}}\sqrt{t}\right)^{1-\alpha} \left(z - 2\sqrt{\frac{1-\alpha}{\alpha}}\sqrt{t}\right)^{\alpha} \tag{2.26}$$

is a solution of the Schwarz–Christoffel equation (2.17) and the backward Loewner evolution (2.23). The corresponding Loewner path is a straight slit starting from the origin growing upward in $\mathbb{H}$ with a tip

$$\gamma(t) = \widehat{f}_t(x_2(t)) = 2\left(\frac{1-\alpha}{\alpha}\right)^{1/2-\alpha} e^{\sqrt{-1}\alpha\pi}\sqrt{t}, \quad t \geq 0. \tag{2.27}$$

Note that the quadratic variation of (2.11) is

$$\langle U, U \rangle_t = \kappa \langle B, B \rangle_t = \kappa t, \quad t \geq 0.$$

It is identified with $x_2(t)^2$, if κ is given by (2.25). SLE will be considered as a randomization of the time-dependent conformal map (2.26).

2.3 Three Phases of SLE

The dependence on D of the Bessel flow given by Theorems 1.1 and 1.2 is mapped to the feature of the SLE$^{(D)}$ paths so that they exhibit three phases.

[Phase 1] When $D \geq D_c = 2$ (i.e., $0 < \kappa \leq \kappa_c \equiv 4$), the SLE$^{(D)}$ path is a *simple curve*, i.e., $\gamma(s) \neq \gamma(t)$ for any $0 \leq s \neq t < \infty$, and $\gamma(0, \infty) \in \mathbb{H}$ (i.e., $\gamma(0, \infty) \cap \mathbb{R} = \emptyset$). In this phase,

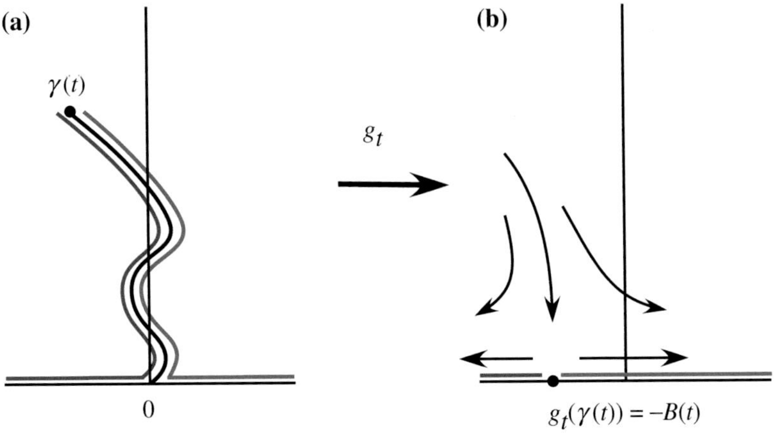

Fig. 2.4 **a** When $D \geq 2$, the SLE$^{(D)}$ path is simple. **b** By g_t, the SLE$^{(D)}$ path is erased from $\mathbb{H}$. The tip of the SLE$^{(D)}$ path, $\gamma(t)$, is mapped to $g_t(\gamma(t)) = -B(t) \in \mathbb{R}$. The flow associated with this conformal transformation is represented by *arrows*

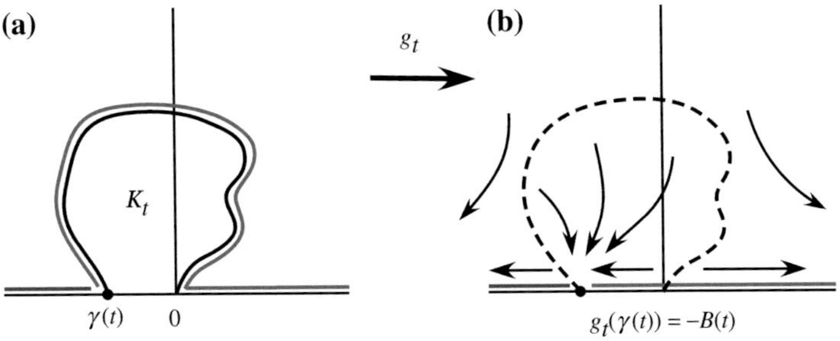

Fig. 2.5 **a** When $3/2 < D < 2$, the SLE$^{(D)}$ path can osculate the real axis. The SLE hull is denoted by K_t. **b** The SLE hull is swallowed. This means that all the points in K_t are simultaneously mapped to a single point $-B(t) \in \mathbb{R}$, which is the image of the tip of SLE$^{(D)}$ path, $\gamma(t)$

$$\mathsf{H}_t = \mathbb{H} \setminus \gamma(0, t], \quad t \geq 0.$$

For each $t \geq 0$, g_t gives a map, which conformally erases a simple curve $\gamma(0, t]$ from $\mathbb{H}$, and the image of the tip $\gamma(t)$ of the SLE path is a BM, $-B(t) \in \mathbb{R} = \partial\mathbb{H}$, as given by (2.15). As shown by Fig. 2.4, it implies that the 'SLE flow' in $\overline{\mathbb{H}}$ is downward in the vertical (imaginary-axis) direction and outward from the position $-B(t)$ in the horizontal (real-axis) direction. Since $Z^z(t) = g_t(z) + B(t)$ by (2.4), if we shift this figure by $B(t)$, we will have a similar picture to Fig. 2.1 for the complexificated version of Bessel flow for $D > 2$.

[Phase 2] When $\overline{D}_{\rm c} = 3/2 < D < D_{\rm c} = 2$ (i.e., $\kappa_{\rm c} = 4 < \kappa < \overline{\kappa}_{\rm c} \equiv 8$), the SLE$^{(D)}$ path can osculate the real axis, $\mathrm{P}(\gamma(0, t] \cap \mathbb{R} \neq \emptyset) > 0, {}^{\forall} t > 0$. Figure 2.5a illustrates the moment $t > 0$ such that the tip of SLE$^{(D)}$ path just

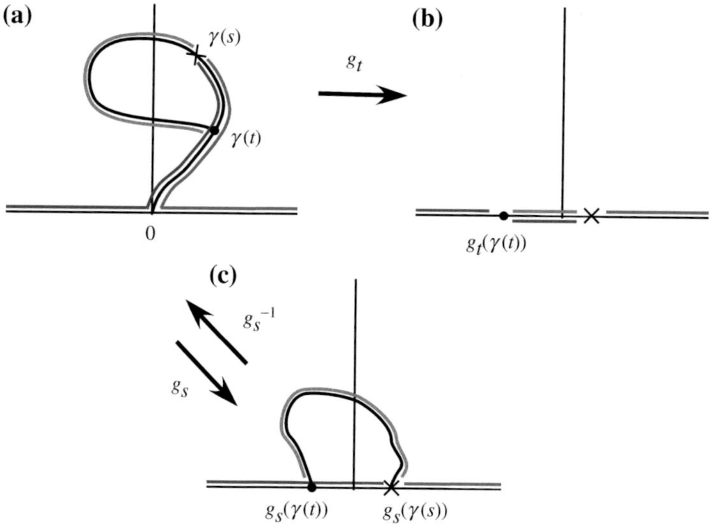

Fig. 2.6 The event that the SLE$^{(D)}$ path osculates $\mathbb{R}$ is equivalent to the event that the SLE$^{(D)}$ path makes a loop

osculates the real axis. The closed region encircled by the path $\gamma(0, t)$ and the line $[\gamma(t), 0] \in \mathbb{R}$ is called an *SLE hull* at time t and denoted by K_t. In this phase

$$\mathsf{H}_t = \mathbb{H} \setminus \mathsf{K}_t, \quad t \geq 0.$$

That is, $g_t(z)$ is a map which erases conformally the SLE hull from $\mathbb{H}$. We can think that by this transformation all the points in K_t are simultaneously mapped to a single point $-B(t) \in \mathbb{R}$, which is the image of the tip $\gamma(t)$. (We say that the hull K_t is *swallowed*. See Fig. 2.5b.) By the definition (2.8), the moment when K_t is swallowed is the time T^z at which the equality $Z^z(t) = g_t(z) + B(t) = 0$ holds $^\forall z \in \mathsf{K}_t$. (Then the RHS of (2.5) diverges and all the points $z \in \mathsf{K}_t$ are lost from the domain of the map g_t.) Theorem 1.2 (ii) states that, when $\overline{D}_c < D < D_c$, two BES$^{(D)}$'s starting from different points $0 < x < y < \infty$ can simultaneously return to the origin. In the complexificated version, all $Z^z(t)$ starting from $z \in \mathsf{K}_t + B(t)$ can arrive at the origin simultaneously (i.e., they are all swallowed).

Osculation of the SLE path with $\mathbb{R}$ means that the SLE path has loops. Figure 2.6a shows the event that the SLE path makes a loop at time $t > 0$. The SLE hull K_t consists of the closed region encircled by the loop and the segment of the SLE path between the origin and the osculating point, and it is completely erased by the conformal transformation g_t from $\mathbb{H}$ as shown by Fig. 2.6b. Let $0 < s < t$ and consider the map g_s, which is the solution of (2.5) at time s. Assume that $\gamma(s)$ is located in the middle of the loop of $\gamma[0, t]$ as shown by Fig. 2.6a. The segment $\gamma[0, s]$ of the SLE path is mapped by g_s to a part of $\mathbb{R}$. Since $\gamma(t)$ osculates a point in $\gamma[0, s]$, its image $g_s(\gamma(t))$ should osculate the real axis $\mathbb{R}$ as shown by Fig. 2.6c.

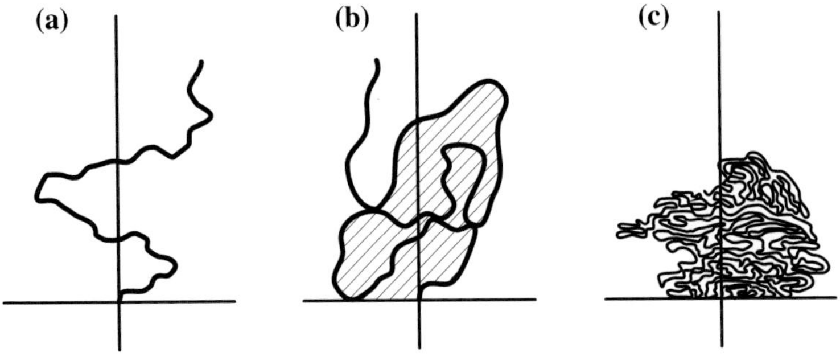

Fig. 2.7 Schematic pictures of $\text{SLE}^{(D)}$ paths in **a** Phase 1 ($D \geq D_c = 2 \Leftrightarrow 0 < \kappa \leq \kappa_c = 4$), **b** Phase 2 ($\overline{D}_c = 3/2 < D < D_c = 2 \Leftrightarrow \kappa_c = 4 < \kappa < \overline{\kappa}_c = 8$), and **c** Phase 3 ($1 < D \leq \overline{D}_c = 3/2 \Leftrightarrow \kappa \geq \overline{\kappa}_c = 8$)

This is the same situation as the one shown in Fig. 2.5a. Since g_s^{-1} is uniquely determined from g_s, the above argument can be reversed. Then the equivalence between the osculation of the SLE path with $\mathbb{R}$ and the self-intersection of the SLE path is concluded.

In this intermediate phase $\overline{D}_c < D < D_c$,

$$\text{SLE}^{(D)} \text{ path } \gamma \text{ is } \textit{self-intersecting, and}$$
$$\bigcup_{t>0} \mathsf{K}_t = \overline{\mathbb{H}} \quad \text{but} \quad \gamma[0, \infty) \cap \mathbb{H} \neq \mathbb{H} \quad \text{with probability 1.}$$

[Phase 3] When $1 < D \leq \overline{D}_c = 3/2$ (i.e., $\kappa \geq \overline{\kappa}_c = 8$), Theorem 1.2 (i) states for the Bessel flow that the ordering $T^x < T^y$ is conserved for any $0 < x < y$. It implies that in this phase the SLE path should be a *space-filling curve*:

$$\gamma[0, \infty) = \overline{\mathbb{H}}.$$

(Otherwise, a swallowing of regions occurs, contradicting Theorem 1.2 (i).)

Figure 2.7 summarizes the three phases of SLE paths.

The SLE paths are fractal curves and their *Hausdorff dimensions* $d_{\text{H}}^{(D)}$ are determined by Beffara [2] as

$$d_{\text{H}}^{(D)} = \begin{cases} 2 & \text{if } 1 < D < \overline{D}_c = \dfrac{3}{2}, \\ \dfrac{2D-1}{2(D-1)} & \text{if } D \geq \overline{D}_c = \dfrac{3}{2}. \end{cases} \tag{2.28}$$

We note that a reciprocity relation is found between D and $d_{\mathrm{H}}^{(D)}$ in **[Phase 1]** and **[Phase 2]**,

$$(D-1)(d_{\mathrm{H}}^{(D)}-1)=\frac{1}{2}, \quad D \geq \overline{D}_{\mathrm{c}}=\frac{3}{2}. \tag{2.29}$$

The stochastic Loewner chain $(g_t)_{t\geq 0}$ as well as the SLE path $\gamma = (\gamma(t))_{t\geq 0}$ are functionals of BM. Therefore for each $D > 1$ we have a statistical ensemble of random curves $\{\gamma(\omega)\}$ in the probability space $(\Omega, \mathscr{F}, \mathrm{P})$ of BM. It is a statistical ensemble of SLE paths $\{\gamma(\omega)\}$ in the upper half plane $\mathbb{H}$, in which they start from the origin: $\gamma(0,\omega)=0$, and approach infinity: $\lim_{t\to\infty}\gamma(t,\omega)=\infty$. We write the probability law of $\{\gamma(\omega)\}$ in such a geometrical setting as $\mathbb{P}_{(\mathbb{H};0,\infty)}$. In general, the probability law of SLE paths $\{\gamma(\omega)\}$ in an simply connected domain $\mathsf{D} \subset \mathbb{C}$, $\mathsf{D} \neq \mathbb{C}$ with $\gamma(0,\omega)=a \in \partial\mathsf{D}$ and $\lim_{t\to\infty}\gamma(t,\omega)=b \in \partial\mathsf{D}$ will be denoted by $\mathbb{P}_{(\mathsf{D};a,b)}$. The important consequence from the facts that BM is a strong Markov process with independent increments and that g_t gives a conformal transformation is the following [13].

(SLE1) The SLE path γ has the following kind of stationary Markov property,

$$\mathbb{P}_{(\mathbb{H};0,\infty)}[\,\cdot\,|\gamma(0,t]] = \mathbb{P}_{(\mathbb{H}\setminus\gamma(0,t];\gamma(t),\infty)}[\,\cdot\,], \quad {}^{\forall}t \geq 0. \tag{2.30}$$

This is called the *domain Markov property*.

(SLE2) Let f be a conformal transformation which maps $\mathbb{H}$ to a domain $\mathsf{D} = f(\mathbb{H})$. Then

$$\mathbb{P}_{(\mathbb{H};0,\infty)}[\,\cdot\,] = \mathbb{P}_{(\mathsf{D};f(0),f(\infty))}[\,\cdot\,]. \tag{2.31}$$

That is, the probability law of γ has *conformal invariance*. Here it should be remarked that the dependence on the geometry of an event should be properly mapped by f. For example, if the event of γ measured by $\mathbb{P}_{(\mathbb{H};0,\infty)}$ depends on a domain $\mathsf{A} \subset \mathbb{H}$, the corresponding event of γ measured by $\mathbb{P}_{(\mathsf{D};f(0),f(\infty))}$ should be considered so that it depends on the domain $f(\mathsf{A})$.

2.4 Cardy's Formula

Let

$$T_{[1,\infty)} = \inf\{t > 0 : \gamma(t) \in [1,\infty)\}.$$

If $D \geq 2$, the SLE$^{(D)}$ path is in **[Phase 1]** and γ does not touch the real axis $\mathbb{R}$ with probability 1. Therefore

$$T_{[1,\infty)} = \infty, \quad D \geq 2 \quad \text{with probability 1}.$$

If $1 < D \leq 3/2$, the SLE$^{(D)}$ path is in **[Phase 3]**, in which γ is a space-filling curve in $\overline{\mathbb{H}}$. Then

$$\gamma(T_{[1,\infty)}) = 1 \quad \text{with probability } 1.$$

When $3/2 < D < 2$, which corresponds to **[Phase 2]**, $\gamma(T_{[1,\infty)})$ has a nontrivial distribution on $[1, \infty)$ as follows.

Proposition 2.1 *Suppose γ is an SLE$^{(D)}$ path with $3/2 < D < 2$. Then, for $x > 0$*

$$\begin{aligned} &\mathbb{P}_{(\mathbb{H};0,\infty)}\Big[\gamma(T_{[1,\infty)}) < 1 + x\Big] \\ &= \frac{\Gamma(D-1)}{\Gamma(2D-3)\Gamma(2-D)} \int_0^{x/(1+x)} \frac{du}{(1-u)^{D-1}u^{2(2-D)}} \qquad (2.32) \\ &= \frac{\Gamma(D-1)}{\Gamma(2(D-1))\Gamma(2-D)} \left(\frac{x}{1+x}\right)^{2D-3} F\left(2D-3, D-1, 2(D-1); \frac{x}{1+x}\right). \qquad (2.33) \end{aligned}$$

Proof By (2.7) and (2.15), we see the equivalence between the events

$$\{\omega : \gamma(T_{[1,\infty)}) < 1 + x\} \iff \{\omega : T^1 < T^{1+x}\}.$$

Then the probability is just obtained from (1.99) and (1.100) in Proposition 1.1 by setting $x \to 1$ and $y \to 1 + x$. Then (2.32) and (2.33) are obtained. □

When $D = 5/3$ (i.e., $\kappa = 6$), (2.32) gives

$$\mathbb{P}_{(\mathbb{H};0,\infty)}\Big[\gamma^{(5/3)}(T_{[1,\infty)}) < 1 + x\Big] = \frac{\Gamma(2/3)}{\Gamma(1/3)^2} \int_0^{x/(1+x)} \frac{du}{(1-u)^{2/3}u^{2/3}}. \qquad (2.34)$$

This formula has the following meaning, and it is called *Cardy's formula* [3, 4].[1]

Let $\triangle$ be a domain in $\mathbb{C}$ whose boundary is the equilateral triangle with vertices $w_1 = 0$, $w_2 = 1$ and $w_3 = e^{\pi\sqrt{-1}/3}$. The conformal map $f_\triangle$ from $\mathbb{H}$ to $\triangle$ satisfying the conditions

$$f_\triangle(0) = w_1, \quad f_\triangle(1) = w_2, \quad f_\triangle(\infty) = w_3 \qquad (2.35)$$

is given by (Exercise 2.3)

$$f_\triangle(z) = \frac{\Gamma(2/3)}{\Gamma(1/3)^2} \int_0^z \frac{du}{u^{2/3}(1-u)^{2/3}}, \quad z \in \mathbb{H}. \qquad (2.36)$$

[1]The original work by Cardy was given on a rectangular domain, but the statement is conformally invariant [3, 4]. The formula becomes particularly easy for an equilateral triangular domain as shown here, and is called *Cardy's formula in Carleson's form* [19]. See Sects. 6.7 and 6.8 in [13].

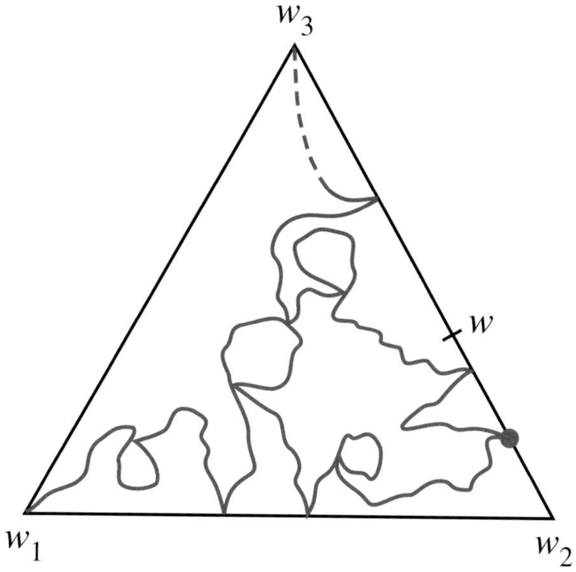

Fig. 2.8 *Red curve* describes an $\mathrm{SLE}^{(5/3)}$ path, $\gamma^{(5/3)} = (\gamma^{(5/3)}(t))_{t\geq 0}$, in $\triangle$ starting from w_1 and approaching w_3 as $t \to \infty$. The point at which $\gamma^{(5/3)}$ first touches the line segment $[w_2, w_3]$ is marked by a *red dot*, which is given by $\gamma^{(5/3)}(T_{[w_2,w_3]})$. In this case $\gamma^{(5/3)}(T_{[w_2,w_3]}) \in [w_2, w]$ for a given $w \in [w_2, w_3]$

We take the branches in the integrand in (2.36) so that (Exercise 2.4)

$$f_\triangle(-x) = e^{\pi\sqrt{-1}/3} \frac{\Gamma(2/3)}{\Gamma(1/3)^2} \int_0^{x/(1+x)} \frac{du}{u^{2/3}(1-u)^{2/3}}, \quad 0 < x < \infty, \tag{2.37}$$

$$f_\triangle(x) = \frac{\Gamma(2/3)}{\Gamma(1/3)^2} \int_0^{x} \frac{du}{u^{2/3}(1-u)^{2/3}}, \quad 0 \leq x \leq 1, \tag{2.38}$$

$$f_\triangle(1+x) = 1 + e^{2\pi\sqrt{-1}/3} \frac{\Gamma(2/3)}{\Gamma(1/3)^2} \int_0^{x/(1+x)} \frac{du}{u^{2/3}(1-u)^{2/3}}, \quad 0 < x < \infty. \tag{2.39}$$

Now we consider the $\mathrm{SLE}^{(5/3)}$ path, $\gamma^{(5/3)}(t)$, $t \in [0, \infty)$ in $\triangle$ starting from w_1 and approaching w_3 as $t \to \infty$. Denote the line segment connecting w_2 and w_3 by $[w_2, w_3]$. Let

$$T_{[w_2,w_3]} = \inf\{t > 0 : \gamma^{(5/3)}(t) \in [w_2, w_3]\}.$$

See Fig. 2.8. For $w \in [w_2, w_3]$, we write the line segment between w_2 and w on $[w_2, w_3]$ as $[w_2, w]$ and the distance between w_2 and w as $|w - w_2|$. By the conformal invariance **(SLE2)** of $\mathrm{SLE}^{(D)}$,

$$\mathbb{P}_{(\triangle;w_1,w_3)}\Big[\gamma^{(5/3)}(T_{[w_2,w_3]}) \in [w_2, f_\triangle(1+x)]\Big] = \mathbb{P}_{(\mathbb{H};0,\infty)}\Big[\gamma^{(5/3)}(T_{[1,\infty)}) < 1+x\Big].$$

Since (2.35) and (2.39) hold, the RHS given by (2.34) is equal to

$$\frac{f_\triangle(1+x)-w_2}{e^{2\pi\sqrt{-1}/3}} = |f_\triangle(1+x)-w_2|.$$

Then we can conclude the following.

Proposition 2.2 *Let $\gamma^{(5/3)}$ be the SLE$^{(5/3)}$ path in $\triangle$ from w_1 to w_3. Then the distribution of $\gamma^{(5/3)}(T_{[w_2,w_3]})$ is uniform on $[w_2, w_3]$. That is,*

$$\mathbb{P}_{(\triangle;w_1,w_3)}\Big[\gamma^{(5/3)}(T_{[w_2,w_3]}) \in [w_2, w]\Big] = |w-w_2| \quad \textit{for any point } w \in [w_2, w_3]. \tag{2.40}$$

2.5 SLE and Statistical Mechanics Models

The highlight of the theory of SLE would be that, if the value of D is properly chosen, the probability law of γ realizes that of the scaling limit of important lattice paths studied in a *statistical mechanics model* exhibiting *critical phenomena* or describing interesting *fractal geometry* defined on an infinite discrete lattice.

The following is a list of the correspondence (up to a conjecture) between the SLE$^{(D)}$ paths with specified values of D, and the names of lattice paths (with the names of models studied in statistical mechanics and fractal physics), whose scaling limits are described by the SLE$^{(D)}$ paths.

SLE$^{(3/2)}$ $\Longleftrightarrow$ random Peano curve (uniform spanning tree) [14]
SLE$^{(5/3)}$ $\Longleftrightarrow$ percolation exploration process (critical percolation model) [19]
SLE$^{(7/4)}$ $\Longleftrightarrow$ FK–Ising interface (critical Ising model) [6, 20]
SLE$^{(2)}$ $\Longleftrightarrow$ random contour curve (Gaussian free surface model) [18]
SLE$^{(7/3)}$ $\Longleftrightarrow$ Ising interface (critical Ising model) [6, 7]
SLE$^{(5/2)}$ $\Longleftrightarrow$ self-avoiding walk [conjecture]
SLE$^{(3)}$ $\Longleftrightarrow$ loop-erased random walk [14]

It is obvious from (2.9) that $D = 3/2, 5/3, 7/4, 2, 7/3, 5/2$, and 3 correspond to $\kappa = 8, 6, 16/3, 4, 3, 8/3$, and 2, respectively.

The SLE$^{(D)}$ path has the following special property if and only if $D = 5/3$: for any $\mathsf{A} \subset \mathbb{H}$ such that $0 \notin \partial\mathsf{A}$, $\infty \notin \partial\mathsf{A}$,

$$\mathbb{P}_{(\mathbb{H};0,\infty)}[\,\cdot\,, \gamma^{(5/3)}(0,t] \cap \mathsf{A} = \emptyset] = \mathbb{P}_{(\mathbb{H}\setminus\mathsf{A};0,\infty)}[\,\cdot\,], \quad t \geq 0.$$

This is called the *locality property*. The lattice path called the *percolation exploration process* $\{\gamma^{\rm per}\}$ defined on the Bernoulli site percolation model [11, 12] studied in statistical mechanics has this property. Cardy conjectured that the scaling limit of $\{\gamma^{\rm per}\}$ obtained from the *critical percolation model* satisfies (2.40) [3]. It was proved by Smirnov [19] for the critical site percolation model on a triangular lattice by showing the conformal invariance of its scaling limit.

The $\text{SLE}^{(D)}$ path has another special property called the *restriction property*, if and only if $D = 5/2$ ($\kappa = 8/3$) [13]: if $\mathsf{D} \subset \mathbb{H}$, $0 \in \partial\mathsf{D}$, $\infty \in \partial\mathsf{D}$, then

$$\mathbb{P}_{(\mathbb{H};0,\infty)}[\,\cdot\,, \gamma^{(5/2)}(0,\infty) \subset \mathsf{D}] = \mathbb{P}_{(\mathsf{D};0,\infty)}[\,\cdot\,].$$

We can see that the *self-avoiding walk* (SAW) [15], which is defined on a lattice and has been studied as a model for polymers, has this property. The conformal invariance of the scaling limit of SAW is, however, not yet proved. If this holds true, then it would imply the equivalence in probability law between the scaling limit of SAW and the $\text{SLE}^{(5/2)}$ path. See [9, 10] for more details and other conjectures.

The relationship between the SLE and the *conformal field theory* (CFT) is discussed in [1, 5]. The *central charge c* and the *scaling dimension h* of the CFT are given as functions of D as

$$c = \frac{(3D-5)(5-2D)}{D-1}, \quad h = \frac{3D-5}{4}, \quad D > 1. \tag{2.41}$$

Exercises

2.1 Derive (2.13) from (2.5).

2.2 Consider the deterministic case where $B(t) \equiv 0$ in (2.5) and

$$\frac{\partial g_t(z)}{\partial t} = \frac{D-1}{2}\frac{1}{g_t(z)}, \quad t \geq 0. \tag{2.42}$$

(i) Solve the equation (2.42) under the initial condition (2.6).
(ii) Determine $\gamma(t)$, $t \geq 0$ by setting $-B(t) \equiv 0$ in (2.15).

2.3 As an application of the Schwarz–Christoffel formula given by Theorem 2.1, prove that (2.36) is the conformal map from $\mathbb{H}$ to Δ satisfying (2.35).

2.4 Derive the expressions (2.37) and (2.39) from (2.36).

References

1. Bauer, M., Bernard, D.: 2D growth processes: SLE and Loewner chains. Phys. Rep. **432**, 115–221 (2006)
2. Beffara, V.: The dimension of the SLE curves. Ann. Probab. **36**, 1421–1452 (2008)
3. Cardy, J.: Critical percolation in finite geometries. J. Phys. A **25**, L201–L206 (1992)
4. Cardy, J.: Lectures on conformal invariance and percolation. (2001). arXiv:math-ph/0103018

5. Cardy, J.: SLE for theoretical physicists. Ann. Phys. **318**, 81–118 (2005)
6. Chelkak, D., Duminil-Copin, H., Hongler, C., Kemppainen, A., Smirnov, S.: Convergence of Ising interfaces to Schramm's SLE curves. C. R. Acad. Sci. Paris, Sér. I Math. **352**, 157–161 (2014)
7. Chelkak, D., Smirnov, S.: Universality in the 2D Ising model and conformal invariance of fermionic observables. Inv. Math. **189**, 515–580 (2012)
8. Driscoll, T.A., Trefethen, L.N.: Schwarz-Christoffel Mapping. Cambridge University Press, New York (2002)
9. Duminil-Copin, H., Smirnov, S.: Conformal invariance in lattice models. In Probability and Statistical Physics in Two and More Dimensions. Proceedings of the Clay Mathematics Institute Summer School and XIV Brazilian School of Probability, Buzios, Brazil, July 11-August 7, 2010, D. Ellwood, C. Newman, V. Sidoravicius, and W. Werner (editors), Clay Math. Inst./Amer. Math. Soc. (2011)
10. Duminil-Copin, H.: Parafermionic Observables and their Applications to Planar Statistical Physics Models. Ensaios Matemáticos 25, pp. 1–371, Brazilian Math. Soc. (2013)
11. Grimmett, G.: Percolation, 2nd edn. Springer, Berlin (1999)
12. Grimmett, G.: The Random-Cluster Model. Springer, Berlin (2006)
13. Lawler, G.F.: Conformally Invariant Processes in the Plane. American Mathematical Society, Providence (2005)
14. Lawler, G.F., Schramm, O., Werner, W.: Conformal invariance of planar loop-erased random walks and uniform spanning trees. Ann. Probab. **32**, 939–995 (2004)
15. Madras, N., Slade, G.: The Self-Avoiding Walk. Birkhäuser, Boston (1993)
16. Rohde, S., Schramm, O.: Basic properties of SLE. Ann. Math. **161**, 883–924 (2005)
17. Schramm, O.: Scaling limits of loop-erased random walks and uniform spanning trees. Israel J. Math. **118**, 221–228 (2000)
18. Schramm, O., Sheffield, S.: The harmonic explorer and its convergence to SLE(4). Ann. Probab. **33**, 2127–2148 (2005)
19. Smirnov, S.: Critical percolation in the plane: conformal invariance, Cardy's formula, scaling limits. C. R. Acad. Sci. Paris, Sér. I Math. **333**, 239–244 (2001)
20. Smirnov, S.: Conformal invariance in random cluster models. I. Holomorphic fermions in the Ising model. Ann. Math. **172**, 1435–1467 (2010)

Chapter 3
Dyson Model

Abstract Dyson's Brownian motion model is a one-parameter ($\beta > 0$) family of interacting Brownian motions in one dimension. A repulsive force acts between any pair of particles, whose strength is given by the inverse of distance of particles multiplied by $\beta/2$. We regard this model as a multivariate extension of the D-dimensional Bessel process with $D = \beta + 1$. We concentrate on the special case where $\beta = 2$ and call it simply the Dyson model, which inherits the following two aspects from the three-dimensional Bessel process: the Dyson model is the eigenvalue process of a Hermitian-matrix-valued Brownian motion (Aspect 1), and it is also constructed as a system of Brownian motions conditioned never to collide with each other (Aspect 2). The notion of determinantal martingale representation is introduced, which leads Aspect 2 into the determinantal property in the sense that all spatio-temporal correlation functions are given by determinants controlled by a single function called the correlation kernel. This strong solvability enables us to construct the Dyson model with an infinite number of particles both in equilibrium and in nonequilibrium. The Tracy–Widom distribution is discussed for the Dyson model.

3.1 Multivariate Extension of Bessel Process

Here we consider the stochastic motion of two particles $(X_1(t), X_2(t))$ in one dimension $\mathbb{R}$ satisfying the following SDEs,

$$\begin{aligned} dX_1(t) &= dB_1(t) + \frac{\beta}{2} \frac{dt}{X_1(t) - X_2(t)}, \\ dX_2(t) &= dB_2(t) + \frac{\beta}{2} \frac{dt}{X_2(t) - X_1(t)}, \end{aligned} \tag{3.1}$$

with the initial condition $x_1 = X_1(0) < x_2 = X_2(0)$ for $0 \le t < \inf\{t > 0 : X_1(t) = X_2(t)\}$, where $B_1(t)$ and $B_2(t)$, $t \ge 0$ are independent BMs and $\beta > 0$ is the 'coupling constant' of the two particles. The second terms in (3.1) represent the repulsive force acting between two particles, which is proportional to the inverse of the distance between them, $X_2(t) - X_1(t)$. Since it is a central force (i.e., depending

M. Katori, *Bessel Processes, Schramm–Loewner Evolution, and the Dyson Model*, SpringerBriefs in Mathematical Physics 11, DOI 10.1007/978-981-10-0275-5_3

only on distance, and thus symmetric for two particles), the 'center of mass' $X_c(t) \equiv (X_2(t) + X_1(t))/2$ is just a time change of BM; we can calculate the quadratic variation as $d\langle X_c, X_c\rangle_t = \langle dX_c, dX_c\rangle_t = \langle (dB_1 + dB_2)/2, (dB_1 + dB_2)/2\rangle_t = dt/2$, since $d\langle B_1, B_1\rangle_t = \langle dB_1, dB_1\rangle_t = dt$, $d\langle B_2, B_2\rangle_t = \langle dB_2, dB_2\rangle_t = dt$ and $d\langle B_1, B_2\rangle_t = \langle dB_1, dB_2\rangle_t = 0$. Then

$$(X_c(t))_{t\ge 0} \overset{\text{(law)}}{=} \left(\frac{1}{\sqrt{2}}B(t)\right)_{t\ge 0} \overset{\text{(law)}}{=} (B(t/2))_{t\ge 0},$$

where $B(t)$ is a BM independent from $B_1(t)$ and $B_2(t)$. On the other hand, if we define the relative coordinate by $X_r(t) \equiv (X_2(t) - X_1(t))/\sqrt{2}$, it satisfies the SDE

$$dX_r(t) = d\widetilde{B}(t) + \frac{\beta}{2}\frac{dt}{X_r(t)} \tag{3.2}$$

for $0 < t < \inf\{t > 0 : X_r(t) = 0\}$, where $\widetilde{B}(t),\ t \ge 0$ is a BM independent from $B_1(t), B_2(t), B(t),\ t \ge 0$ (Exercise 3.1). It is nothing but the SDE for $\text{BES}^{(D)}$ with $D = \beta + 1$:

$$(X_r(t))_{t\ge 0} \overset{\text{(law)}}{=} (R(t))_{t\ge 0}, \quad D = \beta + 1.$$

Dyson [38] introduced N-particle systems of *interacting Brownian motions* in $\mathbb{R}$ as a solution $\mathbf{X}(t) = (X_1(t), X_2(t), \dots, X_N(t))$ of the following system of SDEs: with $\beta > 0$ and the condition $x_1 < x_2 < \cdots < x_N$ for initial positions $x_i = X_i(0), 1 \le i \le N$,

$$dX_i(t) = dB_i(t) + \frac{\beta}{2}\sum_{\substack{1\le j\le N,\\ j\ne i}} \frac{dt}{X_i(t) - X_j(t)}, \quad t \in [0, T^{\mathbf{x}}), \quad 1 \le i \le N, \tag{3.3}$$

where $\{B_i(t)\}_{i=1}^N, t \ge 0$ are independent BMs and

$$T_{ij}^{\mathbf{x}} = \inf\{t > 0 : X_i(t) = X_j(t)\}, \quad 1 \le i < j \le N,$$
$$T^{\mathbf{x}} = \min_{1\le i<j\le N} T_{ij}^{\mathbf{x}}.$$

It is called *Dyson's Brownian motion model with parameter* β [3, 5, 45, 101, 133].

As shown above, in the case where $N = 2$, Dyson's BM model is a composition of a BM (the center of mass) and a $\text{BES}^{(\beta+1)}$ (the relative coordinate). In this sense, Dyson's BM model can be regarded as a multivariate (multidimensional) extension of $\text{BES}^{(\beta+1)}$, $\beta > 0$.

We can prove that, for any $\mathbf{x} \in \mathbb{R}^N$ with $x_1 < x_2 < \cdots < x_N$, $T^{\mathbf{x}} < \infty$ if $\beta < 1$, and $T^{\mathbf{x}} = \infty$ if $\beta \geq 1$ [52, 117]. The critical value $\beta_c = 1$ corresponds to $D_c = \beta_c + 1 = 2$ of $\text{BES}^{(D)}$.[1]

In [6, 7] Dyson's BM model and the related processes are discussed as the special cases of the *Dunkl processes* (see, for instance, [32, 45]).

In particular, we study the special case of Dyson's BM model with parameter $\beta = 2$. We call this special case simply *the Dyson model* [51, 63, 105, 127]. As shown above, the case where $\beta = 2$ corresponds to a $\text{BES}^{(D)}$ with $D = 3$. In Chap. 1, we have shown that $\text{BES}^{(3)}$ has two aspects: **[Aspect 1]** as a radial coordinate of three-dimensional Brownian motion, which was used to define the Bessel process in Sect. 1.8, and **[Aspect 2]** as a one-dimensional Brownian motion conditioned to stay positive as explained in Sect. 1.10. We show that the Dyson model inherits these two aspects from $\text{BES}^{(3)}$ [86].

3.2 Dyson Model as Eigenvalue Process

Dyson introduced the processes (3.3) with $\beta = 1, 2$, and 4 as the eigenvalue processes of matrix-valued stochastic processes in order to realize the point processes in equilibrium called the Gaussian orthogonal ensemble (GOE), the Gaussian unitary ensemble (GUE), and the Gaussian symplectic ensemble (GSE) [3, 5, 38, 45, 101, 133].

Precisely speaking, Dyson considered the Ornstein–Uhlenbeck processes such that as stationary states they realize the eigenvalue distributions of random matrices in GOE, GUE, and GSE. Here we consider matrix-valued Brownian motions instead of the Ornstein–Uhlenbeck processes. Then the variances increase in proportion to time $t \geq 0$. In the following, we will use the stochastic calculus introduced in Chap. 1 in order to explain the Dyson model, but we do not assume any knowledge of *random matrix theory*.

For $\beta = 2$ with given $N \in \mathbb{N}$, we prepare N-tuples of BMs $\{B_{ii}^{x_i}(t)\}_{i=1}^N$, $t \geq 0$, each of which starts from $x_i \in \mathbb{R}$, and $N(N-1)/2$-tuples of pairs of BMs $\{B_{ij}(t), \widetilde{B}_{ij}(t)\}_{1 \leq i < j \leq N}$, $t \geq 0$, starting from the origin. Here, there is a total of $N + 2 \times N(N-1)/2 = N^2$ BMs, each of them independent from the rest. Then consider an $N \times N$ *Hermitian-matrix-valued Brownian motion*,

$$\mathsf{B}^{\mathbf{x}}(t) = \begin{pmatrix} B_{11}^{x_1}(t) & \dfrac{B_{12}(t) + \sqrt{-1}\widetilde{B}_{12}(t)}{\sqrt{2}} & \cdots & \dfrac{B_{1N}(t) + \sqrt{-1}\widetilde{B}_{1N}(t)}{\sqrt{2}} \\ \dfrac{B_{12}(t) - \sqrt{-1}\widetilde{B}_{12}(t)}{\sqrt{2}} & B_{22}^{x_2}(t) & \cdots & \dfrac{B_{2N}(t) + \sqrt{-1}\widetilde{B}_{2N}(t)}{\sqrt{2}} \\ \cdots & \cdots & \cdots & \cdots \\ \dfrac{B_{1N}(t) - \sqrt{-1}\widetilde{B}_{1N}(t)}{\sqrt{2}} & \dfrac{B_{2N}(t) - \sqrt{-1}\widetilde{B}_{2N}(t)}{\sqrt{2}} & \cdots & B_{NN}^{x_N}(t) \end{pmatrix}. \tag{3.4}$$

[1]The existence of a strong and pathwise unique noncolliding solution of SDEs (3.3) for general initial conditions $x_1 \leq x_2 \leq \cdots \leq x_N$ was conjectured by Rogers and Shi [117]. It was proved by Cépa and Lépingle [27] using multivalued SDE theory and by Graczyk and Małecki [53] by classical Itô calculus. See also [33].

Remember that when we introduced $\text{BES}^{(D)}$ in Sect. 1.8, we considered the D-dimensional vector-valued Brownian motion in $\mathbb{R}^N$, (1.56), by preparing D-tuples of independent BMs for its elements. Here we consider the space of $N \times N$ Hermitian matrices denoted by $\mathscr{H}(N)$. Since the dimension of this space is $\dim \mathscr{H}(N) = N^2$, we need N^2 independent BMs for elements to describe a Brownian motion in this space $\mathscr{H}(N)$. Hence we can regard the process $\mathsf{B}^{\mathbf{x}}(t), \ t \geq 0$ defined by (3.4) as a 'Brownian motion in $\mathscr{H}(N)$'. By definition, the initial state of this Brownian motion is the diagonal matrix

$$\mathsf{B}^{\mathbf{x}}(0) = \text{diag}(x_1, x_2, \ldots, x_N). \tag{3.5}$$

We assume $x_1 \leq x_2 \leq \cdots \leq x_N$.

In the usual Gaussian random matrix ensembles, the mean is assumed to be zero. The corresponding matrix-valued BMs are then considered to be started from a zero matrix, i.e., $x_i = 0, 1 \leq i \leq N$ in (3.5). In random matrix theory, the general case where means are non-zero (i.e., $x_i \neq 0$) is discussed with the terminology 'random matrices in an external source' [11, 16, 22, 145]. From the viewpoint of stochastic processes, imposing external sources to break the symmetry of the system corresponds to changing the initial state.

Corresponding to calculating the absolute value (1.57) of $\mathbf{B}^{\mathbf{x}}(t)$, by which $\text{BES}^{(D)}$ was introduced, here we calculate the eigenvalues of $\mathsf{B}^{\mathbf{x}}(t)$. For any $t \geq 0$, there is a family of $N \times N$ unitary matrices $\{\mathsf{U}(t)\}$ which diagonalize $\mathsf{B}^{\mathbf{x}}(t)$,

$$\mathsf{U}^*(t)\mathsf{B}^{\mathbf{x}}(t)\mathsf{U}(t) = \text{diag}(\lambda_1(t), \ldots, \lambda_N(t)) \equiv \Lambda(t), \quad t \geq 0.$$

Here for a matrix $\mathsf{M} = (M_{ij})_{1 \leq i,j \leq N}$, we define its Hermitian conjugate by $\mathsf{M}^* = (\overline{M_{ji}})_{1 \leq i,j \leq N}$, where $\overline{z}$ denotes the complex conjugate of $z \in \mathbb{C}$. Consider a subspace of $\mathbb{R}^N$ defined by

$$\mathbb{W}_N^{\mathsf{A}} \equiv \{\mathbf{x} = (x_1, x_2, \ldots, x_N) \in \mathbb{R}^N : x_1 < x_2 < \cdots < x_N\}, \tag{3.6}$$

which is called the *Weyl chamber of type* A_{N-1} in representation theory. If we impose the condition $(\lambda_i(t))_{i=1}^N \in \mathbb{W}_N^{\mathsf{A}}$, $\mathsf{U}(t)$ is uniquely determined at each time $t \geq 0$.

The main theorem of this section is the following.

Theorem 3.1 *The eigenvalue process $(\lambda_i(t))_{i=1}^N$, $t \geq 0$ of the Hermitian-matrix-valued Brownian motion (3.4) started at (3.5) satisfies the SDEs,*

$$d\lambda_i(t) = dB_i^{x_i}(t) + \sum_{\substack{1 \leq j \leq N, \\ j \neq i}} \frac{dt}{\lambda_i(t) - \lambda_j(t)}, \quad t \geq 0, \quad 1 \leq i \leq N, \tag{3.7}$$

where $(B_i^{x_i}(t))_{i=1}^N$, $t \geq 0$ are independent BMs different from the N^2-tuples of BMs used to define $\mathsf{B}^{\mathbf{x}}(t)$ in (3.4). That is, this process realizes the Dyson model.

The correspondence between $\text{BES}^{(3)}$ and the Dyson model in terms of equivalent processes is summarized as follows.

[Aspect 1]

$\text{BES}^{(3)}$	$\Longleftrightarrow$	radial coordinate of $D = 3$ vector-valued Brownian motion
the Dyson model with N particles	$\Longleftrightarrow$	eigenvalue process of $N \times N$ Hermitian-matrix-valued Brownian motion

Dyson derived (3.7) by applying the perturbation theory in quantum mechanics [38]. Since $(\lambda_i(t))_{i=1}^N$, $t \geq 0$ are functionals of $\{B_{ii}^{x_i}(t), B_{ij}(t), \widetilde{B}_{ij}(t)\}_{1 \leq i < j \leq N}, t \geq 0$, we can use Itô's formula to prove Theorem 3.1. A key point to prove the theorem is applying *Itô's rule for* differentiating the product of *matrix-valued semimartingales* [5, 23, 24, 28, 52, 80, 117, 133]: If $\mathsf{X}(t) = (X_{ij}(t))$ and $\mathsf{Y}(t) = (Y_{ij}(t))$ are $N \times N$ matrices with semimartingale elements, then

$$d(\mathsf{X}^*(t)\mathsf{Y}(t)) = d\mathsf{X}^*(t)\mathsf{Y}(t) + \mathsf{X}^*(t)d\mathsf{Y}(t) + \langle d\mathsf{X}^*, d\mathsf{Y}\rangle_t, \quad t \geq 0, \tag{3.8}$$

where $\langle d\mathsf{X}^*, d\mathsf{Y}\rangle_t$ denotes an $N \times N$ matrix-valued process, whose (i, j)th element is given by the finite-variation process $\sum_k \langle d\overline{X_{ki}}, dY_{kj}\rangle_t$, $1 \leq i, j \leq N$.

In order to demonstrate that Itô's rule for matrix-valued semimartingales is very powerful, we consider the following general setting so that the above result is derived as a special case [80]. Let $\mathsf{H}(t) = (H_{ij}(t))_{1 \leq i,j \leq N}$, $t \geq 0$ be an $N \times N$ Hermitian-matrix-valued diffusion process, where the diagonal elements $H_{ii}^{x_i}(t)$, $1 \leq i \leq N$, $t \geq 0$ are real-valued continuous semimartingales starting from $x_i \in \mathbb{R}$ and the off-diagonal elements $H_{ij}(t)$, $1 \leq i < j \leq N$, $t \geq 0$ are complex-valued continuous semimartingales starting from the origin. The diagonal-matrix-valued process $\Lambda(t) = \text{diag}(\lambda_1(t), \ldots, \lambda_N(t))$ and the unitary-matrix-valued process $\mathsf{U}(t) = (U_{ij}(t))$, $t \geq 0$ are defined so that they satisfy

$$\mathsf{U}^*(t)\mathsf{H}(t)\mathsf{U}(t) = \Lambda(t), \quad t \geq 0. \tag{3.9}$$

Define $\Gamma_{ij}(t)$, $1 \leq i, j \leq N$, by

$$\Gamma_{ij}(t)dt = \big\langle (\mathsf{U}^* d\mathsf{H}\mathsf{U})_{ij}, (\mathsf{U}^* d\mathsf{H}\mathsf{U})_{ji} \big\rangle_t, \tag{3.10}$$

where $d\mathsf{H}(t) = (dH_{ij}(t))_{1 \leq i,j \leq N}$. The finite-variation part of $(\mathsf{U}^*(t)d\mathsf{H}(t)\mathsf{U}(t))_{ii}$, $t \geq 0$ is denoted by $d\Upsilon_i(t)$, $t \geq 0$ for $1 \leq i \leq N$.

Theorem 3.2 *The eigenvalue process* $(\lambda_i(t))_{i=1}^N$, $t \geq 0$ *starting from* $\mathbf{x} \in \mathbb{W}_N^{\text{A}}$ *satisfies the SDEs,*

$$d\lambda_i(t) = dM_i(t) + dJ_i(t), \quad t \geq 0, \quad 1 \leq i \leq N, \tag{3.11}$$

where $(M_i(t))_{1 \leq i \leq N}$, $t \geq 0$ are local martingales with quadratic variations

$$\langle M_i, M_i \rangle_t = \int_0^t \Gamma_{ii}(s)ds, \quad t \geq 0, \quad 1 \leq i \leq N, \tag{3.12}$$

and $(J_i(t))_{1 \leq i \leq N}$, $t \geq 0$ are the finite-variation processes satisfying

$$dJ_i(t) = \sum_{j=1}^N \frac{1}{\lambda_i(t) - \lambda_j(t)} \mathbf{1}_{(\lambda_i(t) \neq \lambda_j(t))} \Gamma_{ij}(t)dt + d\Upsilon_i(t), \quad t \geq 0, \quad 1 \leq i \leq N. \tag{3.13}$$

Before giving the proof of Theorem 3.2, we explain how we can conclude Theorem 3.1 from this general theorem, and give a remark.

Proof of Theorem 3.1. When $\mathsf{H}(t)$, $t \geq 0$ is the Hermitian-matrix-valued Brownian motion, $\mathsf{B}^{\mathbf{x}}(t)$, $t \geq 0$ as given by (3.4), we have

$$\langle dH_{ij}, dH_{k\ell} \rangle_t = \delta_{i\ell} \delta_{jk} dt, \quad 1 \leq i, j, k, \ell \leq N, \quad t \geq 0, \tag{3.14}$$

and (3.10) gives $\Gamma_{ij}(t) \equiv 1, 1 \leq i, j \leq N$ (Exercise 3.2). Moreover, we see that $d\Upsilon_i(t) \equiv 0, t \geq 0, 1 \leq i \leq N$ in this case. Then we obtain (3.7).   □

Remark 3.1   In general, equations (3.11), (3.12) and (3.13) for the eigenvalue process $(\lambda_i(t))_{i=1}^N$, $t \geq 0$ depend on the unitary-matrix-valued process $\mathsf{U}(t)$, $t \geq 0$ through $(\Gamma_{ij}(t))_{1 \leq i,j \leq N}$ and $(\Upsilon_i(t))_{1 \leq i \leq N}$, $t \geq 0$. The equations are written in the form

$$d\lambda_i(t) = \sum_j \alpha_{ij}\left(t, (\lambda_k(t))_{k=1}^N\right)dB_j(t) + \beta_i\left(t, (\lambda_k(t))_{k=1}^N\right)dt, \quad t \geq 0, \quad 1 \leq i \leq N. \tag{3.15}$$

Even if the coefficients α_{ij} and β_i are functions not only of $(\lambda_k(t))_{k=1}^N$ but also of other variables, they are generally called SDEs in [58] (see Definition 1.1 with Eq. (1.1) in Chapter 4). In the special case in which these coefficients only depend on $(\lambda_k(t))_{k=1}^N$, the equations are given in the form

$$d\lambda_i(t) = \sum_j \sigma_{ij}\left((\lambda_k(t))_{k=1}^N\right)dB_j(t) + b_i\left((\lambda_k(t))_{k=1}^N\right)dt, \quad t \geq 0, \quad 1 \leq i \leq N, \tag{3.16}$$

and they are said to be of the *Markovian type* (see Eq. (2.11) in Chap. 4 of [58]). The condition that the SDEs of the eigenvalue processes be reduced to the Markovian type means that the Hermitian-matrix-valued process $\mathsf{H}(t)$, $t \geq 0$ must be unitary invariant in distribution [80]. By virtue of the properties of Brownian motions, the

Hermitian-matrix-valued Brownian motion $\mathsf{B}(t),\ t \geq 0$ defined by (3.4) is unitary invariant in distribution, and thus the obtained SDEs of the Dyson model are of the Markovian type as (3.7).

The rest of this section is devoted to the proof of Theorem 3.2.

Proof of Theorem 3.2. Define a matrix-valued process $\mathsf{A}(t) = (A_{ij}(t))_{1 \leq i,j \leq N}, t \geq 0$ as a solution of the SDE

$$d\mathsf{A}(t) = \mathsf{U}^*(t)d\mathsf{U}(t) + \frac{1}{2}\langle d\mathsf{U}^*, d\mathsf{U}\rangle_t, \quad t \geq 0 \tag{3.17}$$

under the initial condition $\mathsf{A}(0) = 0$. Since $\mathsf{U}^*(t)\mathsf{U}(t) = \mathsf{I}$ for each time t, where I denotes the $N \times N$ unit matrix, Itô's rule (3.8) gives

$$0 = d(\mathsf{U}^*(t)\mathsf{U}(t)) = d\mathsf{U}^*(t)\mathsf{U}(t) + \mathsf{U}^*(t)d\mathsf{U}(t) + \langle d\mathsf{U}^*, d\mathsf{U}\rangle_t.$$

Then the Hermitian conjugate of (3.17) is written as

$$\begin{aligned} d\mathsf{A}^*(t) &= d\mathsf{U}^*(t)\mathsf{U}(t) + \frac{1}{2}\langle d\mathsf{U}^*, d\mathsf{U}\rangle_t \\ &= -\mathsf{U}^*(t)d\mathsf{U}(t) - \frac{1}{2}\langle d\mathsf{U}^*, d\mathsf{U}\rangle_t \\ &= -d\mathsf{A}(t), \end{aligned}$$

that is, $d\mathsf{A}(t)$ is anti-Hermitian matrix-valued. We also see that

$$-\langle d\mathsf{A}, d\mathsf{A}\rangle_t = \langle d\mathsf{U}^*\mathsf{U}, \mathsf{U}^*d\mathsf{U}\rangle_t = \langle d\mathsf{U}^*, d\mathsf{U}\rangle_t, \tag{3.18}$$

since $\langle d\mathsf{U}^*\mathsf{U}, \mathsf{U}^*d\mathsf{U}\rangle_t = \langle d\mathsf{U}^*, \mathsf{U}\mathsf{U}^*d\mathsf{U}\rangle_t$ and $\mathsf{U}(t)\mathsf{U}^*(t) = \mathsf{I}$. Then by multiplying $\mathsf{U}(t)$ from the left to (3.17), we have

$$d\mathsf{U}(t) = \mathsf{U}(t)\left(d\mathsf{A}(t) + \frac{1}{2}\langle d\mathsf{A}, d\mathsf{A}\rangle_t\right). \tag{3.19}$$

Applying Itô's rule (3.8) to (3.9), we have

$$\begin{aligned} d\Lambda(t) &= d\mathsf{U}^*(t)\mathsf{H}(t)\mathsf{U}(t) + \mathsf{U}^*(t)d\mathsf{H}(t)\mathsf{U}(t) + \mathsf{U}^*(t)\mathsf{H}(t)d\mathsf{U}(t) \\ &\quad + \langle d\mathsf{U}^*, d\mathsf{H}\mathsf{U}\rangle_t + \langle d\mathsf{U}^*, \mathsf{H}d\mathsf{U}\rangle_t + \langle \mathsf{U}^*d\mathsf{H}, d\mathsf{U}\rangle_t \\ &= \mathsf{U}^*(t)d\mathsf{H}(t)\mathsf{U}(t) + \left\{\Lambda(t)\mathsf{U}^*(t)d\mathsf{U}(t) + (\Lambda(t)\mathsf{U}^*(t)d\mathsf{U}(t))^*\right\} \\ &\quad + \left\{\langle \mathsf{U}^*d\mathsf{H}, d\mathsf{U}\rangle_t + \langle \mathsf{U}^*d\mathsf{H}, d\mathsf{U}\rangle_t^*\right\} + \langle d\mathsf{U}^*, \mathsf{H}d\mathsf{U}\rangle_t. \end{aligned}$$

Using (3.19), the terms in the RHS are rewritten as

$$\Lambda(t)\mathsf{U}^*(t)d\mathsf{U}(t) = \Lambda(t)d\mathsf{A}(t) + \frac{1}{2}\Lambda(t)\langle d\mathsf{A}, d\mathsf{A}\rangle_t,$$
$$\langle \mathsf{U}^* d\mathsf{H}, d\mathsf{U}\rangle_t = \langle \mathsf{U}^* d\mathsf{H}, \mathsf{U} d\mathsf{A}\rangle_t,$$

and

$$\langle d\mathsf{U}^*, \mathsf{H} d\mathsf{U}\rangle_t = \langle d\mathsf{A}^*, \mathsf{U}^*\mathsf{H}\mathsf{U} d\mathsf{A}\rangle_t = \langle d\mathsf{A}^*, \Lambda d\mathsf{A}\rangle_t.$$

Then we have the equality

$$\begin{aligned} d\Lambda(t) &= \mathsf{U}^*(t)d\mathsf{H}(t)\mathsf{U}(t) + \Lambda(t)d\mathsf{A}(t) + (\Lambda(t)d\mathsf{A}(t))^* \\ &\quad + \frac{1}{2}\Lambda(t)\langle d\mathsf{A}, d\mathsf{A}\rangle_t + \frac{1}{2}(\Lambda(t)\langle d\mathsf{A}, d\mathsf{A}\rangle_t)^* \\ &\quad + \langle \mathsf{U}^* d\mathsf{H}, \mathsf{U} d\mathsf{A}\rangle_t + \langle \mathsf{U}^* d\mathsf{H}, \mathsf{U} d\mathsf{A}\rangle_t^* + \langle d\mathsf{A}^*, \Lambda d\mathsf{A}\rangle_t. \end{aligned} \tag{3.20}$$

The ith diagonal element of (3.20) gives

$$\begin{aligned} d\lambda_i(t) &= \sum_{k,\ell} \overline{U_{ki}(t)} U_{\ell i}(t) dH_{k\ell}(t) \\ &\quad + 2\lambda_i(t) d\gamma_{ii}(t) + d\phi_{ii}(t) + \overline{d\phi_{ii}(t)} + d\psi_{ii}(t), \quad 1 \le i \le N, \end{aligned} \tag{3.21}$$

and the (i, j)th off-diagonal element of (3.20) gives

$$\begin{aligned} 0 &= \sum_{k,\ell} \overline{U_{ki}(t)} U_{\ell j}(t) dH_{k\ell}(t) + \lambda_i(t) dA_{ij}(t) + \lambda_j(t) d\overline{A_{ji}(t)} + \lambda_i(t) d\gamma_{ij}(t) \\ &\quad + \lambda_j(t) d\overline{\gamma_{ji}(t)} + d\phi_{ij}(t) + d\overline{\phi_{ji}(t)} + d\psi_{ij}(t), \quad 1 \le i < j \le N, \end{aligned} \tag{3.22}$$

where we have used the abbreviations

$$\begin{aligned} d\gamma_{ij}(t) &\equiv \frac{1}{2}(\langle d\mathsf{A}, d\mathsf{A}\rangle_t)_{ij} = \frac{1}{2}\sum_k \langle dA_{ik}, dA_{kj}\rangle_t = d\overline{\gamma_{ji}(t)}, \\ d\phi_{ij}(t) &\equiv (\langle \mathsf{U}^* d\mathsf{H}, \mathsf{U} d\mathsf{A}\rangle_t)_{ij} = \sum_{k,\ell,m} \overline{U_{ki}(t)} U_{\ell m}(t) \langle dH_{k\ell}, dA_{mj}\rangle_t, \\ d\psi_{ij}(t) &\equiv (\langle d\mathsf{A}^*, \Lambda d\mathsf{A}\rangle_t)_{ij} \\ &= \sum_k \lambda_k(t) \langle d\overline{A_{ki}}, dA_{kj}\rangle_t = -\sum_k \lambda_k(t) \langle dA_{ik}, dA_{kj}\rangle_t. \end{aligned} \tag{3.23}$$

Since $(\gamma_{ij}(t))_{1 \le i,j \le N}$, $(\phi_{ij}(t))_{1 \le i,j \le N}$, $(\psi_{ij}(t))_{1 \le i,j \le N}$, $t \ge 0$, are finite-variation processes, (3.21) gives

$$d\langle M_i, M_i\rangle_t \equiv \langle d\lambda_i, d\lambda_i\rangle_t = \big\langle (\mathsf{U}^* d\mathsf{H}\mathsf{U})_{ii}, (\mathsf{U}^* d\mathsf{H}\mathsf{U})_{ii}\big\rangle_t$$
$$= \sum_{k,\ell}\sum_{m,n} \overline{U_{ki}(t)} U_{\ell i}(t) \overline{U_{mi}(t)} U_{ni}(t) \langle dH_{k\ell}, dH_{mn}\rangle_t.$$

By the definition (3.10) of $\Gamma_{ij}(t)$, this proves (3.12).

On the other hand, since $d\mathsf{A}(t)$ is anti-Hermitian-valued, (3.22) gives

$$(\lambda_j(t) - \lambda_i(t)) dA_{ij}(t) = \sum_{k,\ell} \overline{U_{ki}(t)} U_{\ell j}(t) dH_{k\ell}(t)$$
$$+ (\lambda_i(t) + \lambda_j(t)) d\gamma_{ij}(t) + d\phi_{ij}(t) + d\overline{\phi_{ji}(t)} + d\psi_{ij}(t).$$

This implies

$$\sum_{k,\ell} \overline{U_{ki}(t)} U_{\ell j}(t) dH_{k\ell}(t) = (\lambda_j(t) - \lambda_i(t)) dA_{ij}(t) + (\text{finite-variation processes}), \tag{3.24}$$

and we can rewrite (3.23) as

$$d\phi_{ij}(t) = \sum_{k,\ell,m} \overline{U_{ki}(t)} U_{\ell m}(t) \langle dH_{k\ell}, dA_{mj}\rangle_t$$
$$= \sum_k (\lambda_k(t) - \lambda_i(t)) \langle dA_{ik}, dA_{kj}\rangle_t.$$

Then the second line of the RHS of (3.21) is written as

$$2\lambda_i(t) d\gamma_{ii}(t) + d\phi_{ii}(t) + d\overline{\phi_{ii}(t)} + d\psi_{ii}(t)$$
$$= \sum_j \Big\{ \lambda_i(t) + 2(\lambda_j(t) - \lambda_i(t)) - \lambda_j(t) \Big\} \langle dA_{ij}, dA_{ji}\rangle_t$$
$$= \sum_j (\lambda_j(t) - \lambda_i(t)) \langle dA_{ij}, dA_{ji}\rangle_t$$
$$= \sum_j \frac{1}{\lambda_i(t) - \lambda_j(t)} \mathbf{1}_{(\lambda_i(t) \neq \lambda_j(t))} \sum_{k,\ell,m,n} \overline{U_{ki}(t)} U_{\ell j}(t) \overline{U_{mj}(t)} U_{ni}(t) \langle dH_{k\ell}, dH_{mn}\rangle_t,$$

where (3.24) was used in the last equality. By the definitions of $\Gamma_{ij}(t)$ and $d\Upsilon_i(t)$, the finite-variation part of (3.21) is equal to (3.13). The proof is completed. □

3.3 Dyson Model as Noncolliding Brownian Motion

In this section, we extend the formula (1.77) for $\text{BES}^{(3)}$ to the Dyson model.

First we rewrite (1.78) as follows. Consider a set of two operations, identity ($\sigma_1 =$ id) and reflection ($\sigma_2 =$ ref), such that for $x \in \mathbb{R}$, $\sigma_1(x) = x$ and $\sigma_2(x) = -x$, and signatures are given as $\text{sgn}(\sigma_1) = 1$ and $\text{sgn}(\sigma_2) = -1$, respectively. Then we have

$$\begin{aligned} q^{\text{abs}}(t, y|x) &= \sum_{\sigma \in \{\text{id},\text{ref}\}} \text{sgn}(\sigma) p(t, y|\sigma(x)) \\ &= \sum_{\sigma \in \{\text{id},\text{ref}\}} \text{sgn}(\sigma) p(t, \sigma(y)|x), \quad x, y \in \mathbb{R}, \quad t \geq 0. \end{aligned} \tag{3.25}$$

Then we consider the set of all permutations of N indices $\{1, 2, \ldots, N\}$, which is denoted by $\mathscr{S}_N$, and put the following multivariate function

$$\sum_{\sigma \in \mathscr{S}_N} \text{sgn}(\sigma) \prod_{i=1}^{N} p(t, y_{\sigma(i)}|x_i) = \det_{1 \leq i, j \leq N} [p(t, y_i|x_j)] \tag{3.26}$$

of $\mathbf{x} = (x_1, \ldots, x_N) \in \mathbb{W}_N^{\text{A}}$ and $\mathbf{y} = (y_1, \ldots, y_N) \in \mathbb{W}_N^{\text{A}}$ with a parameter $t \geq 0$.

As a multidimensional extension of the absorbing Brownian motion in $\mathbb{R}_+$ with an absorbing wall at $x = 0$, we consider the *absorbing Brownian motion* $\mathbf{B}^{\mathbf{x}}(t) = (B_1^{x_1}(t), \ldots, B_N^{x_N}(t))$ in $\mathbb{W}_N^{\text{A}}$. The starting point $\mathbf{x}$ is assumed to be in $\mathbb{W}_N^{\text{A}}$. We put absorbing walls at the boundaries of $\mathbb{W}_N^{\text{A}}$. When $\mathbf{B}^{\mathbf{x}}$ hits any of the walls, it is absorbed and the process is stopped. In other words, the Brownian motion $\mathbf{B}^{\mathbf{x}}$ started at $\mathbf{x} \in \mathbb{W}_N^{\text{A}}$ is killed when it arrives at the boundaries of $\mathbb{W}_N^{\text{A}}$. We define $q_N(t, \mathbf{y}|\mathbf{x})$ for $\mathbf{x}, \mathbf{y} \in \mathbb{W}_N^{\text{A}}$, $t \geq 0$ as the probability density for the event such that this absorbing Brownian motion starting from $\mathbf{x}$ at time $t = 0$ 'survives' up to time t and arrives at $\mathbf{y}$ at the time t. Note that the boundaries of $\mathbb{W}_N^{\text{A}}$ are the hyperplanes $x_i = x_j$, $1 \leq i < j \leq N$ in $\mathbb{R}^N$. Then, if we interpret $\mathbf{x} \in \mathbb{R}^N$ as a configuration of N particles on a line $\mathbb{R}$, this absorbing Brownian motion in $\mathbb{W}_N^{\text{A}}$ can be regarded as an N-particle system such that each particle executes BM when distances between neighboring particles are positive, but when any two neighboring particles collide, the process is stopped. This process is a continuum limit (diffusion scaling limit) [78, 79] of the *vicious walker model* on $\mathbb{Z}$ introduced by Fisher [43] (see also [9, 10, 26, 30, 39, 41, 64, 74, 91]).

The following is known as the *Karlin–McGregor formula* [67]. See also [61, 88, 121]. Note that the discrete analogue is known as the *Lindström–Gessel–Viennot formula* [50, 93].

Lemma 3.1 *The transition probability density of the absorbing Brownian motion in* $\mathbb{W}_N^{\text{A}}$ *is given by (3.26). That is,*

$$q_N(t, \mathbf{y}|\mathbf{x}) = \det_{1 \leq i, j \leq N} [p(t, y_i|x_j)], \quad \mathbf{x}, \mathbf{y} \in \mathbb{W}_N^{\text{A}}, \quad t \geq 0. \tag{3.27}$$

Proof By the property **(BM2)** and the definition of the transition probability density of BM, $p(t, y|x)$ gives the total probability mass of the Brownian path $\pi[0, t]$ from x to y with time duration t. Let $\Pi_t(x, y)$ denote the collection of all Brownian paths from $x \in \mathbb{R}$ to $y \in \mathbb{R}$ with time duration $t \geq 0$. We will interpret (3.26) as a generating function for $(N+1)$-tuples, $(\sigma, \pi_1, \ldots, \pi_N)$, where $\sigma \in \mathscr{S}_N$, $\pi_i = \pi_i[0,t] \in \Pi_t(x_i, y_{\sigma(i)})$, $1 \leq i \leq N$. Under the assumption $\mathbf{x} \in \mathbb{W}_N^{\mathrm{A}}$, let

$$\tau = \inf\{t > 0 : \mathbf{B}^{\mathbf{x}}(t) \notin \mathbb{W}_N^{\mathrm{A}}\}. \tag{3.28}$$

Assume that $\tau < t$ and $B_k^{x_k}(\tau) = B_\ell^{x_\ell}(\tau)$. For a pair of paths (π_k, π_ℓ), we define (π'_k, π'_ℓ) by exchanging the Brownian paths of π_k, π_ℓ after $t = \tau$:

$$\pi'_k[0,t] = \pi_k[0,\tau] \cup \pi_\ell(\tau, t], \quad \pi'_\ell[0,t] = \pi_\ell[0,\tau] \cup \pi_k(\tau, t].$$

We define $\pi'_i = \pi_i$ for $i \neq k, \ell$ and $\sigma' = \sigma \circ \sigma_{k\ell}$, where $\sigma_{k\ell}$ denotes the exchange of k and ℓ. Then the operation $(\sigma, \pi_1, \ldots, \pi_N) \mapsto (\sigma', \pi'_1, \ldots, \pi'_N)$ is an involution. By this operation, the absolute value of the contribution to the generating function (3.26) is not changed because of the strong Markov property (1.8) and the reflection principle of BM (1.5), but the sign is changed. So the contribution of any such pairs $\{(\sigma, \pi_1, \ldots, \pi_N), (\sigma', \pi'_1, \ldots, \pi'_N)\}$ is canceled out. The remaining non-zero contributions in (3.26) are from N-tuples of nonintersecting Brownian paths. Since $\mathbf{x}, \mathbf{y} \in \mathbb{W}_N^{\mathrm{A}}$, $\sigma = \mathrm{id}$ and so $\mathrm{sgn}(\sigma) = \mathrm{sgn}(\mathrm{id}) = 1$ for nonintersecting paths. Hence (3.26) gives the total probability mass of N-tuples of nonintersecting Brownian paths from $\mathbf{x}$ to $\mathbf{y}$ with time duration t and is identified with $q_N(t, \mathbf{y}|\mathbf{x})$ for the absorbing Brownian motion in $\mathbb{W}_N^{\mathrm{A}}$. □

For an initial configuration $\mathbf{x} \in \mathbb{W}_N^{\mathrm{A}}$, the survival probability of the absorbing Brownian motion in $\mathbb{W}_N^{\mathrm{A}}$ up to time $t \geq 0$ is then given by

$$\mathrm{P}^{\mathbf{x}}[\tau > t] = \int_{\mathbb{W}_N^{\mathrm{A}}} q_N(t, \mathbf{y}|\mathbf{x}) d\mathbf{y}, \quad t \geq 0, \tag{3.29}$$

where $d\mathbf{y} = \prod_{i=1}^N dy_i$.

Now we consider an N-variate extension of $\mathrm{BES}^{(3)}$, which is the N-particle system of BMs in $\mathbb{R}$ conditioned never to collide with each other, that is, they do not collide even during the time interval (t, ∞). We simply call this process the *noncolliding Brownian motion* with N particles. The *transition probability density function* $p_N(t, \mathbf{y}|\mathbf{x})$ of this process from $\mathbf{x} \in \mathbb{W}_N^{\mathrm{A}}$ to $\mathbf{y} \in \mathbb{W}_N^{\mathrm{A}}$ with time duration $t \geq 0$ should be obtained by the following limit (cf. (1.127) in Exercise 1.14),

$$p_N(t, \mathbf{y}|\mathbf{x}) = \lim_{T \to \infty} \frac{q_N(t, \mathbf{y}|\mathbf{x}) \mathrm{P}^{\mathbf{y}}[\tau > T - t]}{\mathrm{P}^{\mathbf{x}}[\tau > T]}. \tag{3.30}$$

Let

$$h_N(\mathbf{x}) = \prod_{1 \leq i < j \leq N} (x_j - x_i). \tag{3.31}$$

Then the following are proved.

Proposition 3.1 (i) *The transition probability density of the noncolliding Brownian motion with N particles is given by*

$$p_N(t, \mathbf{y}|\mathbf{x}) = \frac{h_N(\mathbf{y})}{h_N(\mathbf{x})} q_N(t, \mathbf{y}|\mathbf{x}), \quad \mathbf{x}, \mathbf{y} \in \mathbb{W}_N^{\mathrm{A}}, \quad t \geq 0, \tag{3.32}$$

where q_N is the Karlin–McGregor determinant (3.27).

(ii) *Let $|\mathbf{x}| = \sqrt{\sum_{i=1}^N x_i^2}$. Then*

$$\begin{aligned} p_N(t, \mathbf{y}|\mathbf{0}) &\equiv \lim_{|\mathbf{x}| \to 0} p_N(t, \mathbf{y}|\mathbf{x}) \\ &= \frac{t^{-N(N-1)/2}}{\prod_{n=1}^{N-1} n!} h_N(\mathbf{y})^2 \prod_{i=1}^N p(t, y_i|0), \quad \mathbf{y} \in \mathbb{W}_N^{\mathrm{A}}, \quad t \geq 0. \end{aligned} \tag{3.33}$$

Before giving the proof of this proposition, we will present the main statement in this section [51].

Theorem 3.3 *The noncolliding Brownian motion is equivalent in probability law with the Dyson model.*

Proof Denote the N-dimensional Laplacian with respect to the variables $\mathbf{x} = (x_1, \ldots, x_N)$ by $\Delta^{(N)} \equiv \sum_{i=1}^N \partial^2/\partial x_i^2$. Provided $\mathbf{x}, \mathbf{y} \in \mathbb{W}_N^{\mathrm{A}}$, we can verify that (3.32) satisfies the following partial differential equation (PDE),

$$\frac{\partial}{\partial t} p_N(t, \mathbf{y}|\mathbf{x}) = \frac{1}{2} \Delta^{(N)} p_N(t, \mathbf{y}|\mathbf{x}) + \sum_{\substack{1 \leq i, j \leq N, \\ i \neq j}} \frac{1}{x_i - x_j} \frac{\partial}{\partial x_i} p_N(t, \mathbf{y}|\mathbf{x}) \tag{3.34}$$

with the initial condition $p_N(0, \mathbf{y}|\mathbf{x}) = \delta(\mathbf{y} - \mathbf{x}) \equiv \prod_{i=1}^N \delta(y_i - x_i)$ (Exercise 3.3). It can be regarded as the backward Kolmogorov equation of the stochastic process with N particles, $\mathbf{X}(t) = (X_1(t), \ldots, X_N(t))$, which solves the system of SDEs,

$$dX_i(t) = dB_i^{x_i}(t) + \sum_{\substack{1 \leq j \leq N, \\ j \neq i}} \frac{dt}{X_i(t) - X_j(t)}, \quad t \geq 0, \quad 1 \leq i \leq N. \tag{3.35}$$

Equation (3.35) is identified with the case where $\beta = 2$ in (3.3). Then the equivalence between the Dyson model and the noncolliding Brownian motion is proved [51]. □

As already noted in Sect. 1.10, $h_1(x) \equiv x$ is a positive harmonic function in $\mathbb{R}_+ = (0, \infty)$ which satisfies the condition $h_1(0) = 0$. Similarly, we can see that

$$\Delta^{(N)} h_N(\mathbf{x}) = 0, \tag{3.36}$$

and

$$h_N(\mathbf{x}) > 0, \text{ if } \mathbf{x} \in \mathbb{W}_N^{\mathrm{A}}, \quad \text{and} \quad h_N(\mathbf{x}) = 0, \text{ if } \mathbf{x} \in \partial \mathbb{W}_N^{\mathrm{A}}. \tag{3.37}$$

Proposition 3.1 (i) states that the noncolliding Brownian motion is the h-transformation of the absorbing Brownian motion in $\mathbb{W}_N^{\mathrm{A}}$, in which the harmonic function is given by (3.31) [34, 51].

The formula (3.33) given in Proposition 3.1 (ii) can be regarded as a multivariate extension of $p^{(3)}(t, y|0)$ given by (1.125) in Exercise 1.13. The transition probability density (3.33) determines an *entrance law* from the configuration $\mathbf{0}$ (i.e., the state with all particles at the origin) for the Dyson model (3.35), which will be discussed in Sect. 3.8 (see (3.105)). See also the footnote on p. 59.

The analogy of **[Aspect 2]** between $\mathrm{BES}^{(3)}$ and the Dyson model is summarized as follows.

[**Aspect 2**]

$\mathrm{BES}^{(3)}$	$\Longleftrightarrow$	h-transformation of absorbing BM in $\mathbb{R}_+$
	$\Longleftrightarrow$	BM conditioned to stay positive
the Dyson model	$\Longleftrightarrow$	h-transformation of absorbing BM in $\mathbb{W}_N^{\mathrm{A}}$
	$\Longleftrightarrow$	noncolliding BM

In the rest of this section, we will explain how to prove Proposition 3.1. By the multilinearity of determinants, (3.27) with (1.1) gives (Exercise 3.4)

$$q_N(t, \mathbf{y}|\mathbf{x}) = \frac{1}{(2\pi t)^{N/2}} e^{-(|\mathbf{x}|^2+|\mathbf{y}|^2)/2t} \det_{1\le i,j\le N} [e^{y_i x_j/t}]. \tag{3.38}$$

The following Maclaurin expansion is well-known,

$$e^{y_i x_j/t} = \sum_{n=0}^{\infty} \left(\frac{y_i x_j}{t}\right)^n \frac{1}{n!}. \tag{3.39}$$

Here we want to expand the multivariate function $\det_{1\le i,j\le N}[e^{y_i x_j/t}]$ by polynomials of $\{x_i\}_{i=1}^N$ and $\{y_i\}_{i=1}^N$. A useful basis of symmetric polynomials in $\{x_i\}_{i\ge1}$ is given by *Schur functions* [48, 95, 128]. They are labeled by *partitions* $\mu = (\mu_1, \mu_2, \dots)$, which are sets of nonnegative integers in decreasing order $\mu_1 \ge \mu_2 \ge \cdots$. The nonzero μ_i's in a partition μ are called *parts* of μ, and the number of parts is called *length* of μ and denoted by $\ell(\mu)$. The Schur polynomial of $\{x_i\}_{i=1}^N$ labeled by μ with $\ell(\mu) = N \in \mathbb{N}_0$ is defined by

$$s_\mu(\mathbf{x}) = \frac{\det_{1\le i,j\le N} [x_j^{\mu_i+N-i}]}{\det_{1\le i,j\le N} [x_j^{N-i}]}. \tag{3.40}$$

Here the denominator is especially called the *Vandermonde determinant*, which is equal to the *product of differences* (Exercise 3.5),

$$V_N(\mathbf{x}) \equiv \det_{1\le i,j\le N}[x_j^{N-i}] = \prod_{1\le i<j\le N}(x_i - x_j).$$

It is easy to see that $V_N(\mathbf{x}) = (-1)^{N(N-1)/2} h_N(\mathbf{x})$. We can show that the ratio of two determinants in (3.40) indeed gives a symmetric polynomial of $\{x_i\}_{i=1}^N$ (Exercises 3.6 and 3.7). By the definition (3.40), we see that for $\mathbf{0} = (0, 0, ...)$,

$$s_\mu(\mathbf{0}) = \begin{cases} 1, & \text{if } \mu = \emptyset, \\ 0, & \text{otherwise,} \end{cases} \tag{3.41}$$

where the null partition is denoted by $\emptyset = (0, 0, \ldots)$.

Now we prove the following. Here $\mathbf{x}/\sqrt{t} = (x_1/\sqrt{t}, \ldots, x_N/\sqrt{t})$.

Lemma 3.2 *For* $\mathbf{x}, \mathbf{y} \in \mathbb{R}^N$,

$$\begin{aligned} &\det_{1\le i,j\le N}[e^{x_i y_j/t}] \\ &= h_N(\mathbf{x}/\sqrt{t})h_N(\mathbf{y}/\sqrt{t}) \sum_{\mu,\ell(\mu)\le N} \frac{1}{\prod_{i=1}^N(\mu_i + N - i)!} s_\mu(\mathbf{x}/\sqrt{t})s_\mu(\mathbf{y}/\sqrt{t}) \\ &= \frac{t^{-N(N-1)/2}}{\prod_{n=1}^{N-1} n!} h_N(\mathbf{x})h_N(\mathbf{y}) \times \left\{1 + \mathcal{O}\left(\frac{|\mathbf{x}|}{\sqrt{t}}\right)\right\} \quad \textit{as } \frac{|\mathbf{x}|}{\sqrt{t}} \to 0. \end{aligned} \tag{3.42}$$

Proof By the Maclaurin expansion (3.39) and the multilinearity of determinants, we have

$$\begin{aligned} \det_{1\le i,j\le N}[e^{x_i y_j/t}] &= \det_{1\le i,j\le N}\left[\sum_{n=0}^\infty \left(\frac{x_i y_j}{t}\right)^n \frac{1}{n!}\right] \\ &= \sum_{\mathbf{n}=(n_1,n_2,\cdots,n_N)\in\mathbb{N}_0^N} \prod_{k=1}^N \frac{1}{n_k!} \det_{1\le i,j\le N}\left[\left(\frac{x_i y_j}{t}\right)^{n_i}\right]. \end{aligned} \tag{3.43}$$

We can see that for any symmetric function $f(\mathbf{n})$ of $\mathbf{n} \in \mathbb{N}_0^N$,

$$\sum_{\mathbf{n}\in\mathbb{N}_0^N} f(\mathbf{n}) \det_{1\le i,j\le N}\left[\left(\frac{x_i y_j}{t}\right)^{n_i}\right] = \sum_{\mathbf{n}\in\mathbb{N}_0^N} f(\mathbf{n}) \frac{1}{N!} \sum_{\sigma\in\mathcal{S}_N} \det_{1\le i,j\le N}\left[\left(\frac{x_i y_j}{t}\right)^{n_{\sigma(i)}}\right], \tag{3.44}$$

and we can prove that (Exercise 3.8)

$$\sum_{\sigma\in\mathcal{S}_N} \det_{1\le i,j\le N}\left[\left(\frac{x_i y_j}{t}\right)^{n_{\sigma(i)}}\right] = \det_{1\le i,j\le N}\left[\left(\frac{x_i}{\sqrt{t}}\right)^{n_j}\right] \det_{1\le k,\ell\le N}\left[\left(\frac{y_k}{\sqrt{t}}\right)^{n_\ell}\right]. \tag{3.45}$$

Since $\det_{1\le i,j\le N}[(x_i/\sqrt{t})^{n_j}] = 0$ if $n_{j_1} = n_{j_2}$ for any pair $1 \le j_1 \ne j_2 \le N$, (3.43) equals

$$\sum_{\mathbf{n}\in\mathbb{N}_0^N, 0\le n_1<n_2<\cdots<n_N} \prod_{m=1}^{N} \frac{1}{n_m!} \det_{1\le i,j\le N}\left[\left(\frac{x_i}{\sqrt{t}}\right)^{n_j}\right] \det_{1\le k,\ell\le N}\left[\left(\frac{y_k}{\sqrt{t}}\right)^{n_\ell}\right].$$

Now we change the variables in the summation from $\{n_i\}$ to $\{\mu_i\}$ by $\mu_i = n_i - N + i$, $1 \le i \le N$. Using (3.40) we obtain the first line of (3.42). By (3.41), the estimation in $|\mathbf{x}|/\sqrt{t} \to 0$ is given as the second line of (3.42). □

By this lemma, we have the following asymptotics,

$$q_N(t, \mathbf{y}|\mathbf{x}) = \frac{1}{C_N} t^{-N^2/2} h_N(\mathbf{x}) h_N(\mathbf{y}) e^{-|\mathbf{y}|^2/2t} \times \left\{1 + \mathcal{O}\left(\frac{|\mathbf{x}|}{\sqrt{t}}\right)\right\} \quad \text{as } \frac{|\mathbf{x}|}{\sqrt{t}} \to 0 \tag{3.46}$$

with $C_N = (2\pi)^{N/2} \prod_{n=1}^{N-1} n!$. The integral formula for $a > 0, \gamma > 0$,

$$\int_{\mathbb{R}^N} e^{-a|\mathbf{x}|^2} |h_N(\mathbf{x})|^{2\gamma} d\mathbf{x} = (2\pi)^{N/2}(2a)^{-N(\gamma(N-1)+1)/2} \prod_{i=1}^{N} \frac{\Gamma(1+i\gamma)}{\Gamma(1+\gamma)}, \tag{3.47}$$

is found in [101] (Eq. (17.6.7), p. 321) as a variation of the *Selberg integral* [122], whose proof was given in [94]. Using the special case of this Selberg–Mehta–Macdonald equality, we obtain the following.

Lemma 3.3 *Let $C'_N = 2^{N/2} \prod_{i=1}^{N} \Gamma(i/2)$. Then for $\mathbf{x} \in \mathbb{W}_N^{\mathrm{A}}$, $t > 0$,*

$$\mathrm{P}^{\mathbf{x}}[\tau > t] = \frac{C'_N}{C_N} t^{-N(N-1)/4} h_N(\mathbf{x}) \times \left\{1 + \mathcal{O}\left(\frac{|\mathbf{x}|}{\sqrt{t}}\right)\right\} \quad as\ \frac{|\mathbf{x}|}{\sqrt{t}} \to 0. \tag{3.48}$$

Proof If we set in (3.47) $\gamma = 1/2$, $a = 1/2t$ and note that the integral over $\mathbb{R}^N$ can be replaced by the integral over $\mathbb{W}_N^{\mathrm{A}}$ multiplied by $N!$, since the integrand is symmetric in $\mathbf{x}$, we have

$$\int_{\mathbb{W}_N^{\mathrm{A}}} e^{-|\mathbf{x}|^2/2t} h_N(\mathbf{x}) d\mathbf{x} = C'_N t^{N(N+1)/4}. \tag{3.49}$$

Then (3.29) with (3.46) proves the lemma. □

Proof of Proposition 3.1. (i) Fix $\mathbf{x}, \mathbf{y} \in \mathbb{W}_N^{\mathrm{A}}$ and $t \in [0, \infty)$. Then Lemma 3.3 proves that (3.30) is given by (3.32).
(ii) The asymptotics (3.46) of q_N implies

$$\lim_{|\mathbf{x}|\to 0} p_N(t, \mathbf{y}|\mathbf{x}) = \frac{t^{-N^2/2}}{C_N} h_N(\mathbf{y})^2 e^{-|\mathbf{y}|^2/2t}.$$

This gives the expression (3.33). □

We can apply the argument given in this section also for processes associated with Weyl chambers of other types [51, 80, 89]. See [37, 75] for the noncolliding Brownian motion associated with the Weyl alcoves.

3.4 Determinantal Martingale Representation (DMR)

As **[Aspect 2]**, the Dyson model is constructed as the h-transformation of the absorbing Brownian motion in $\mathbb{W}_N^{\mathrm{A}}$. Therefore, at any positive time $t > 0$ the configuration is an element of $\mathbb{W}_N^{\mathrm{A}}$,

$$\mathbf{X}(t) = (X_1(t), X_2(t), \ldots, X_N(t)) \in \mathbb{W}_N^{\mathrm{A}}, \quad t > 0, \tag{3.50}$$

and hence there are no multiple points at which coincidence of particle positions, $X_i(t) = X_j(t), i \neq j$, occurs. We can consider, however, the Dyson model starting from initial configurations with multiple points. In order to describe configurations with multiple points, we represent each particle configuration by a sum of delta measures in the form

$$\xi(\cdot) = \sum_{i\in\mathbb{I}} \delta_{x_i}(\cdot) \tag{3.51}$$

with a sequence of points in $\mathbb{R}$, $\mathbf{x} = (x_i)_{i\in\mathbb{I}}$, where $\mathbb{I}$ is a countable index set. Here for $y \in \mathbb{R}$, $\delta_y(\cdot)$ denotes the delta measure such that $\delta_y(\{x\}) = 1$ for $x = y$ and $\delta_y(\{x\}) = 0$ otherwise. Then, for (3.51) and $A \subset \mathbb{R}$, $\xi(A) = \int_A \xi(dx) = \sum_{i\in\mathbb{I}:x_i\in A} 1 = \sharp\{x_i, x_i \in A\}$.

If the total number of particles N is finite, then $\mathbb{I} = \{1, 2, \ldots, N\}$, but we would like to also consider the cases where $N = \infty$ later. The measures of the form (3.51) satisfying the condition $\xi(K) < \infty$ for any compact subset $K \subset \mathbb{R}$ are called the *nonnegative integer-valued Radon measures* on $\mathbb{R}$ and we denote the space they form by $\mathfrak{M}$. The set of configurations without multiple points is denoted by $\mathfrak{M}_0 = \{\xi \in \mathfrak{M} : \xi(\{x\}) \leq 1, {}^\forall x \in \mathbb{R}\}$. There is a trivial correspondence between $\mathbb{W}_N^{\mathrm{A}}$ and $\mathfrak{M}_0$. We call $\mathbf{x} \in \mathbb{R}^N$ a *labeled configuration* and $\xi \in \mathfrak{M}$ an *unlabeled configuration*.

We introduce a sequence of independent BMs, $\mathbf{B}^{\mathbf{x}}(t) = (B_i^{x_i}(t))_{i\in\mathbb{I}}$, $t \geq 0$, in $(\Omega, \mathscr{F}, \mathrm{P}^{\mathbf{x}})$ with expectation written as $\mathrm{E}^{\mathbf{x}}$.

In this section we assume that $\xi = \sum_{i\in\mathbb{I}} \delta_{x_i} \in \mathfrak{M}_0$, $\xi(\mathbb{R}) = N \in \mathbb{N}$ and consider the Dyson model as an $\mathfrak{M}_0$-valued diffusion process,

$$\Xi(t, \cdot) = \sum_{i=1}^{N} \delta_{X_i(t)}(\cdot), \quad t \geq 0, \tag{3.52}$$

starting from the initial configuration $\xi = \sum_{i=1}^{N} \delta_{x_i}$, where $\mathbf{X}(t) = (X_1(t), \cdots, X_N(t))$ is the solution of (3.35) under the initial configuration $\mathbf{x} = (x_1, \ldots, x_N) \in \mathbb{W}_N^{\mathrm{A}}$. We write the process as $(\Xi, \mathbb{P}^{\xi})$ and express the expectation with respect to the probability law $\mathbb{P}^{\xi}$ of the Dyson model by $\mathbb{E}^{\xi}[\,\cdot\,]$. We introduce a filtration $\{(\mathscr{F}_{\Xi})_t\}_{t \in [0,\infty)}$ on the space of continuous paths $\mathrm{C}([0, \infty) \to \mathfrak{M})$ defined by $(\mathscr{F}_{\Xi})_t = \sigma(\Xi(s), s \in [0, t])$, where σ denotes the smallest σ-field.

[Aspect 2] of the Dyson model is expressed by the following equality: for any $(\mathscr{F}_{\Xi})_t$-measurable bounded function F, $0 \leq t \leq T < \infty$,

$$\mathbb{E}^{\xi}[F(\Xi(\cdot))] = \mathrm{E}^{\mathbf{x}}\left[F\left(\sum_{i=1}^{N} \delta_{B_i(\cdot)}\right) \mathbf{1}_{(\tau > T)} \frac{h_N(\mathbf{B}(T))}{h_N(\mathbf{x})}\right], \tag{3.53}$$

where τ is defined by (3.28) and we have assumed the relations $\xi = \sum_{i=1}^{N} \delta_{x_i} \in \mathfrak{M}_0$, $\mathbf{x} = (x_1, \ldots, x_N) \in \mathbb{W}_N^{\mathrm{A}}$ and (3.52).

In the following lemma, we claim that even if we delete the indicator $\mathbf{1}_{(\tau > T)}$ in the RHS of (3.53), still the equality holds. It is a multivariate extension of the claim by which we replaced (1.79) by (1.80) in Sect. 1.10.

Lemma 3.4 *Assume that* $\xi = \sum_{i=1}^{N} \delta_{x_i} \in \mathfrak{M}_0$, $\mathbf{x} = (x_1, \ldots, x_N) \in \mathbb{W}_N^{\mathrm{A}}$. *For any* $(\mathscr{F}_{\Xi})_t$*-measurable bounded function* F, $0 \leq t \leq T < \infty$,

$$\mathbb{E}^{\xi}[F(\Xi(\cdot))] = \mathrm{E}^{\mathbf{x}}\left[F\left(\sum_{i=1}^{N} \delta_{B_i(\cdot)}\right) \frac{h_N(\mathbf{B}(T))}{h_N(\mathbf{x})}\right]. \tag{3.54}$$

Proof It is sufficient to consider the case where F is given as

$$F(\Xi(\cdot)) = \prod_{m=1}^{M} g_m(\mathbf{X}(t_m)) \tag{3.55}$$

for an arbitrary $M \in \mathbb{N}$, and an arbitrary subdivision $\Delta_M([0, T])$ with $0 \equiv t_0 < t_1 < \cdots < t_{M-1} < t_M \equiv T < \infty$ with bounded measurable functions g_m on $\mathbb{R}^N$, $1 \leq m \leq M$. Since the particles are unlabeled in the process $(\Xi, \mathbb{P}^{\xi})$, g_m's are symmetric functions. Now we consider the following specified form for (3.53):

$$\mathbb{E}^{\xi}\left[\prod_{m=1}^{M} g_m(\mathbf{X}(t_m))\right] = \mathrm{E}^{\mathbf{x}}\left[\mathbf{1}_{(\tau > t_M)} \prod_{m=1}^{M} g_m(\mathbf{B}(t_m)) \frac{h_N(\mathbf{B}(t_M))}{h_N(\mathbf{x})}\right]. \tag{3.56}$$

We introduce the stopping times

$$\tau_{ij} = \inf\{t > 0 : B_i(t) = B_j(t)\}, \quad 1 \le i < j \le N.$$

Let $\sigma_{ij} \in \mathscr{S}_N$ be the permutation of (i, j), $1 \le i, j \le N$. For $\mathbf{x} = (x_k)_{k=1}^N$, we write $\sigma_{ij}(\mathbf{x}) = \mathbf{y} = (y_k)_{k=1}^N$ such that $y_i = x_j$, $y_j = x_i$, and $y_k = x_k$, $k \ne i, k \ne j$. Note that in a configuration $\mathbf{y}$, if $y_i = y_j$, $i \ne j$, then $\sigma_{ij}(\mathbf{y}) = \mathbf{y}$, and the processes $\mathbf{B}$ and $\sigma_{ij}(\mathbf{B})$ are identical in distribution under the probability measure $\mathrm{P}^{\mathbf{y}}$. By the strong Markov property of the process $\mathbf{B}(t)$, $t \ge 0$ and by the fact that h_N is anti-symmetric, while g_m, $1 \le m \le M$ are symmetric,

$$\mathrm{E}^{\mathbf{x}}\left[\mathbf{1}_{(\tau=\tau_{ij}<t_M)} \prod_{m=1}^{M} g_m(\mathbf{B}(t_m)) \frac{h_N(\mathbf{B}(t_M))}{h_N(\mathbf{x})}\right] = 0, \quad 1 \le i < j \le N.$$

Since $\mathrm{P}^{\mathbf{x}}(\tau_{ij} = \tau_{i'j'}) = 0$ if $(i, j) \ne (i', j')$, and $\tau = \min_{1 \le i < j \le N} \tau_{ij}$,

$$\mathrm{E}^{\mathbf{x}}\left[\mathbf{1}_{(\tau<t_M)} \prod_{m=1}^{M} g_m(\mathbf{B}(t_m)) \frac{h_N(\mathbf{B}(t_M))}{h_N(\mathbf{x})}\right] = 0.$$

Hence, (3.56) gives the equality

$$\mathbb{E}^{\xi}\left[\prod_{m=1}^{M} g_m(\mathbf{X}(t_m))\right] = \mathrm{E}^{\mathbf{x}}\left[\prod_{m=1}^{M} g_m(\mathbf{B}(t_m)) \frac{h_N(\mathbf{B}(t_M))}{h_N(\mathbf{x})}\right].$$

The statement is proved. □

In Sect. 1.2, we introduced the fundamental martingale polynomials associated with BM, $\{m_n(t, x)\}_{n \in \mathbb{N}_0}$. Since they are monic polynomials, we see that (Exercise 3.9)

$$\begin{aligned} \frac{h_N(\mathbf{y})}{h_N(\mathbf{x})} &= \frac{1}{h_N(\mathbf{x})} \det_{1 \le i, j \le N} [y_j^{i-1}] \\ &= \frac{1}{h_N(\mathbf{x})} \det_{1 \le i, j \le N} [m_{i-1}(t, y_j)] \end{aligned} \tag{3.57}$$

for an arbitrary $t \in [0, \infty)$. This implies that $(h_N(\mathbf{B}(t))/h_N(\mathbf{x}))_{t \ge 0}$ is a martingale.

Here we extend the integral transformation defined by (1.25) with (1.22) in Sect. 1.2 to a linear integral transformation of multivariate functions as follows. When $F^{(i)}(\mathbf{x}) = \prod_{j=1}^N f_j^{(i)}(x_j)$, $i = 1, 2$ are given for $\mathbf{x} = (x_1, \ldots, x_N) \in \mathbb{R}^N$, then

$$\mathscr{I}\left[F^{(i)}(\mathbf{W}) \left| \{(t_\ell, x_\ell)\}_{\ell=1}^N \right.\right] = \prod_{j=1}^{N} \mathscr{I}\left[f_j^{(i)}(W_j) \middle| (t_j, x_j)\right], \quad i = 1, 2,$$

and

$$\mathscr{I}\Big[c_1 F^{(1)}(\mathbf{W}) + c_2 F^{(2)}(\mathbf{W})\,\Big|\{(t_\ell, x_\ell)\}_{\ell=1}^N\Big]$$
$$= c_1 \mathscr{I}\Big[F^{(1)}(\mathbf{W})\,\Big|\{(t_\ell, x_\ell)\}_{\ell=1}^N\Big] + c_2 \mathscr{I}\Big[F^{(2)}(\mathbf{W})\,\Big|\{(t_\ell, x_\ell)\}_{\ell=1}^N\Big],$$

$c_1, c_2 \in \mathbb{C}$, for $0 < t_i < \infty, 1 \leq i \leq N$, where $\mathbf{W} = (W_1, \dots, W_N) \in \mathbb{R}^N$. In particular, if $t_\ell = t$, $1 \leq {}^\forall \ell \leq N$, we write $\mathscr{I}[\cdot|\{(t_\ell, x_\ell)\}_{\ell=1}^N]$ simply as $\mathscr{I}[\cdot|(t, \mathbf{x})]$ with $\mathbf{x} = (x_1, \dots, x_N)$. Then (3.57) is further rewritten as

$$\begin{aligned}
\frac{h_N(\mathbf{y})}{h_N(\mathbf{x})} &= \frac{1}{h_N(\mathbf{x})} \det_{1 \leq i,j \leq N} \Big[\mathscr{I}[(W_j)^{i-1}|(t, y_j)]\Big] \\
&= \mathscr{I}\left[\frac{1}{h_N(\mathbf{x})} \det_{1 \leq i,j \leq N}[(W_j)^{i-1}]\,\middle|\,(t, \mathbf{y})\right] \\
&= \mathscr{I}\left[\frac{h_N(\mathbf{W})}{h_N(\mathbf{x})}\,\middle|\,(t, \mathbf{y})\right],
\end{aligned} \tag{3.58}$$

where the multilinearity of determinants has been used.

Now we use the following determinant identity [73, 87, 90].

Lemma 3.5 *For* $\mathbf{x} = (x_1, \dots, x_N) \in \mathbb{W}_N^{\mathrm{A}}$, $\mathbf{z} = (z_1, \dots, z_N) \in \mathbb{C}^N$,

$$\frac{h_N(\mathbf{z})}{h_N(\mathbf{x})} = \det_{1 \leq i,j \leq N} \Big[\Phi_\xi^{x_i}(z_j)\Big], \tag{3.59}$$

where

$$\Phi_\xi^u(z) = \prod_{\substack{1 \leq k \leq N, \\ x_k \neq u}} \frac{z - x_k}{u - x_k} \tag{3.60}$$

for $\xi = \sum_{i=1}^N \delta_{x_i} \in \mathfrak{M}_0$ $z, u \in \mathbb{C}$.

Proof Let

$$H(\mathbf{x}, \mathbf{z}) = \det_{1 \leq i,j \leq N} \left[\prod_{1 \leq k \leq N, k \neq i} (z_j - x_k)\right].$$

It is a polynomial function of x_i, z_i, $1 \leq i \leq N$ with degree $N(N-1)$ satisfying the conditions that $H(\mathbf{x}, \mathbf{x}) = (-1)^{N(N-1)/2} h_N(\mathbf{x})^2$, and $H(\mathbf{x}, \mathbf{z}) = 0$, if $x_i = x_j$ or $z_i = z_j$ for some i, j with $1 \leq i < j \leq N$. Hence we have $H(\mathbf{x}, \mathbf{z}) = (-1)^{N(N-1)/2} h_N(\mathbf{x}) h_N(\mathbf{z})$. By the definition (3.60), the RHS of (3.59) is equal to $H(\mathbf{x}, \mathbf{z})/(-1)^{N(N-1)/2} h_N(\mathbf{x})^2$ for $\xi \in \mathfrak{M}_0$, and then we obtain (3.59). □

Then (3.58) is written as

$$\frac{h_N(\mathbf{y})}{h_N(\mathbf{x})} = \mathscr{I}\left[\det_{1\leq i,j\leq N}[\Phi_\xi^{x_i}(W_j)]\Big|(t,\mathbf{y})\right]$$
$$= \det_{1\leq i,j\leq N}[\mathscr{M}_\xi^{x_i}(t,y_j)], \tag{3.61}$$

where

$$\mathscr{M}_\xi^x(t,y) = \mathscr{I}[\Phi_\xi^x(W)|(t,y)], \quad x,y\in\mathbb{R}, \quad t\geq 0. \tag{3.62}$$

Proposition 3.2 *Assume that* $\xi = \sum_{i=1}^N \delta_{x_i} \in \mathfrak{M}_0$. *The following are satisfied by (3.62).*

(i) $(\mathscr{M}_\xi^{x_i}(t,B(t)))_{t\geq 0}$, $1\leq i\leq N$ *are continuous martingales.*
(ii) *For any time* $t\geq 0$, $\mathscr{M}_\xi^{x_i}(t,y)$, $1\leq i\leq N$ *are linearly independent functions of* y.
(iii) *For* $1\leq i,j\leq N$, $\lim_{t\downarrow 0}\mathrm{E}^{x_i}[\mathscr{M}_\xi^{x_j}(t,B(t))] = \delta_{ij}$.

Then for any $(\mathscr{F}_\Xi)_t$*-measurable bounded function* F, $0\leq t\leq T<\infty$, *the equality*

$$\mathbb{E}^\xi[F(\Xi(\cdot))] = \mathrm{E}^{\mathbf{x}}\left[F\left(\sum_{i=1}^N \delta_{B_i(\cdot)}\right)\mathscr{D}_\xi(T,\mathbf{B}(T))\right] \tag{3.63}$$

holds, where

$$\mathscr{D}_\xi(t,\mathbf{y}) = \det_{1\leq i,j\leq N}[\mathscr{M}_\xi^{y_j}(t,y_i)], \quad \mathbf{y}=(y_1,\ldots,y_N)\in\mathbb{W}_N^{\mathrm{A}}, \quad t\geq 0. \tag{3.64}$$

Proof For each $1\leq i\leq N$, $\Phi_\xi^{x_i}(z)$ is a polynomial of z with degree $N-1$. Then it can be written as $\Phi_\xi^{x_i}(z) = \sum_{n=0}^{N-1} a_n z^n$, where the coefficients $a_n\in\mathbb{R}$ are functions of ξ (i.e., x_j, $1\leq j\leq N$). By the definition (3.62) and (1.26) in Sect. 1.2, $\mathscr{M}_\xi^{x_i}(t,x) = \sum_{n=0}^{N-1} a_n m_n(t,x)$, where the functions $\{m_n(t,x)\}_{n\in\mathbb{N}_0}$ are the fundamental martingale polynomials given by (1.18). Then (i) follows from Lemma 1.1 (iii). Since $\xi\in\mathfrak{M}_0$ is assumed, the set of zeroes of $\Phi_\xi^{x_i}(z)$ is different from that of $\Phi_\xi^{x_j}(z)$, if $i\neq j$, and the condition (ii) is satisfied. By (3.60), $\Phi_\xi^{x_j}(x_i) = \delta_{ij}$, $1\leq i,j\leq N$. Thus the condition (iii) is also satisfied. The equality (3.63) with (3.64) is obtained from (3.54) of Lemma 3.4 together with (3.61). □

We remark that $\mathscr{D}_\xi(t,\mathbf{B}(t))$, $t\geq 0$ is indeed a continuous martingale by part (i) and is not identically zero by part (ii) of Proposition 3.2. We call $\mathscr{D}_\xi(t,\mathbf{B}(t))$, $t\geq 0$ a *determinantal martingale* and the equality (3.63) the *determinantal martingale representation* (DMR) of noncolliding Brownian motion [73].

Since $\Phi_\xi^u(z)$ given by (3.60) is a polynomial of z, the argument given in Sect. 1.7 provides

$$\mathscr{M}_\xi^u(t, B(t)) = \widetilde{\mathrm{E}}[\Phi_\xi^u(Z(t))], \quad t \geq 0, \tag{3.65}$$

where $Z(t),\ t \geq 0$ is a complex Brownian motion defined by (1.54) and $\widetilde{\mathrm{E}}$ denotes the expectation with respect to the imaginary part of $Z(t)$, $\Im Z(t) = \widetilde{B}(t)$.

For each $B_i^{x_i}(t)$, $i \in \mathbb{I}$, we introduce an independent one-dimensional BM starting from the origin, $\widetilde{B}_i(t)$, and define a complex Brownian motion as $Z_i(t) = B_i(t) + \sqrt{-1}\widetilde{B}_i(t)$, $i \in \mathbb{I}$. We write the expectation with respect to $\{\widetilde{B}_i(t)\}_{i\in\mathbb{I}}$ as $\widetilde{\mathrm{E}}$ and define $\mathbf{E}^{\mathbf{x}} = \mathrm{E}^{\mathbf{x}} \otimes \widetilde{\mathrm{E}}$. Then the DMR (3.63) is rewritten as follows.

Lemma 3.6 *Assume that* $\xi = \sum_{i=1}^N \delta_{x_i} \in \mathfrak{M}_0$. *For any* $(\mathscr{F}_\Xi)_t$*-measurable bounded function* F, $0 \leq t \leq T < \infty$, *the following equality holds,*

$$\mathbb{E}^\xi[F(\Xi(\cdot))] = \mathbf{E}^{\mathbf{x}}\left[F\left(\sum_{i=1}^N \delta_{\Re Z_i(\cdot)}\right) \det_{1\leq i,j\leq N}\left[\Phi_\xi^{x_i}(Z_j(T))\right]\right]. \tag{3.66}$$

This is a special case of DMR [73] and it was called the *complex Brownian motion representation* in [87].

As we have seen in Sect. 1.7, the complex Brownian motion is conformal invariant, and each $\Phi_\xi^x(Z_j(t))$, $x \in \mathbb{R}$ is a time change of a complex Brownian motion, $Z_j(\cdot)$. Then the expectation is conserved,

$$\mathbf{E}^x[\Phi_\xi^x(Z_j(t))] = \mathbf{E}^x[\Phi_\xi^x(Z_j(T))], \quad 0 \leq^\forall t \leq T < \infty, \tag{3.67}$$

for $x \in \mathbb{R}$. That is, for $x \in \mathbb{R}$, $\{\Phi_\xi^x(Z_j(t))\}_{1\leq j\leq N}$ are independent *conformal local martingales* (see, for example, Sect. 5.2 of [116]).

Lemma 3.6 means that any observables of the Dyson model are calculated by a system of independent complex Brownian motions, whose paths are weighted by a multivariate complex function $\det_{1\leq i,j\leq N}[\Phi_\xi^{x_i}(Z_j(T))]$, which is a conformal local martingale.

3.5 Reducibility of DMR and Correlation Functions

For $n \in \mathbb{N}$, an index set $\{1, 2, \dots, n\}$ is denoted by $\mathbb{I}_n$. Fixing $N \in \mathbb{N}$ with $N' \in \mathbb{I}_N$, we write $\mathbb{J} \subset \mathbb{I}_N$, $\sharp\mathbb{J} = N'$, if $\mathbb{J} = \{j_1, \dots, j_{N'}\}$, $1 \leq j_1 < \cdots < j_{N'} \leq N$. For $\mathbf{x} = (x_1, \dots, x_N) \in \mathbb{R}^N$, put $\mathbf{x}_\mathbb{J} = (x_{j_1}, \dots, x_{j_{N'}})$. In particular, we write $\mathbf{x}_{N'} = \mathbf{x}_{\mathbb{I}_{N'}}$, $1 \leq N' \leq N$. (By definition $\mathbf{x}_N = \mathbf{x}$.) A collection of all permutations of elements in $\mathbb{J}$ is denoted by $\mathscr{S}(\mathbb{J})$. In particular, we write $\mathscr{S}_{N'} = \mathscr{S}(\mathbb{I}_{N'})$, $1 \leq N' \leq N$.

The following shows the *reducibility* of the determinantal martingale in the sense that, if we observe a symmetric function depending on N' variables, $N' \leq N$, then the size of determinantal martingale can be reduced from N to N'.

Lemma 3.7 *Assume that* $\xi = \sum_{i=1}^{N} \delta_{x_i} = \sum_{i \in \mathbb{I}_N} \delta_{x_i}$ *with* $\mathbf{x} \in \mathbb{W}_N^{\mathrm{A}}$. *Let* $1 \leq N' \leq N$. *For* $0 < t \leq T < \infty$ *and an* $(\mathscr{F}_{\Xi})_t$*-measurable symmetric function* $F_{N'}$ *on* $\mathbb{R}^{N'}$,

$$\sum_{\mathbb{J} \subset \mathbb{I}_N, \sharp \mathbb{J} = N'} \mathrm{E}^{\mathbf{x}} \left[F_{N'}(\mathbf{B}_{\mathbb{J}}(t)) \mathscr{D}_{\xi}(T, \mathbf{B}(T)) \right]$$
$$= \int_{\mathbb{W}_{N'}^{\mathrm{A}}} \xi^{\otimes N'}(d\mathbf{v}) \mathrm{E}^{\mathbf{v}} \left[F_{N'}(\mathbf{B}_{N'}(t)) \mathscr{D}_{\xi}(T, \mathbf{B}_{N'}(T)) \right]. \tag{3.68}$$

Proof By the definition (3.64) the LHS of (3.68) is equal to

$$\sum_{\mathbb{J} \subset \mathbb{I}_N, \sharp \mathbb{J} = N'} \mathrm{E}^{\mathbf{x}} \left[F_{N'}(\mathbf{B}_{\mathbb{J}}(t)) \det_{i,j \in \mathbb{I}_N} [\mathscr{M}_{\xi}^{x_j}(T, B_i(T))] \right]$$
$$= \sum_{\mathbb{J} \subset \mathbb{I}_N, \sharp \mathbb{J} = N'} \mathrm{E}^{\mathbf{x}} \left[F_{N'}(\mathbf{B}_{\mathbb{J}}(t)) \sum_{\sigma \in \mathscr{S}_N} \mathrm{sgn}(\sigma) \prod_{i=1}^{N} \mathscr{M}_{\xi}^{x_{\sigma(i)}}(T, B_i(T)) \right]$$
$$= \sum_{\mathbb{J} \subset \mathbb{I}_N, \sharp \mathbb{J} = N'} \sum_{\sigma \in \mathscr{S}_N} \mathrm{sgn}(\sigma)$$
$$\times \mathrm{E}^{\mathbf{x}} \left[F_{N'}(\mathbf{B}_{\mathbb{J}}(t)) \prod_{i \in \mathbb{J}} \mathscr{M}_{\xi}^{x_{\sigma(i)}}(T, B_i(T)) \prod_{j \in \mathbb{I}_N \setminus \mathbb{J}} \mathscr{M}_{\xi}^{x_{\sigma(j)}}(T, B_j(T)) \right]$$

Since $(B_i(t))_{t \geq 0}$, $1 \leq i \leq N$ are independent, it is equal to

$$\sum_{\mathbb{J} \subset \mathbb{I}_N, \sharp \mathbb{J} = N'} \sum_{\sigma \in \mathscr{S}_N} \mathrm{sgn}(\sigma) \mathrm{E}^{\mathbf{x}} \left[F_{N'}(\mathbf{B}_{\mathbb{J}}(t)) \prod_{i \in \mathbb{J}} \mathscr{M}_{\xi}^{x_{\sigma(i)}}(T, B_i(T)) \right]$$
$$\times \prod_{j \in \mathbb{I}_N \setminus \mathbb{J}} \mathrm{E}^{\mathbf{x}} \left[\mathscr{M}_{\xi}^{x_{\sigma(j)}}(T, B_j(T)) \right]. \tag{3.69}$$

By the property (i) of $\mathscr{M}_{\xi}$ in Proposition 3.2,

$$\prod_{j \in \mathbb{I}_N \setminus \mathbb{J}} \mathrm{E}^{\mathbf{x}} \left[\mathscr{M}_{\xi}^{x_{\sigma(j)}}(T, B_j(T)) \right] = \prod_{j \in \mathbb{I}_N \setminus \mathbb{J}} \mathrm{E}^{\mathbf{x}} \left[\mathscr{M}_{\xi}^{x_{\sigma(j)}}(t, B_j(t)) \right], \quad {}^{\forall} t \in [0, T],$$

and by the property (iii) of $\mathscr{M}_{\xi}$ in Proposition 3.2, this is equal to $\prod_{j \in \mathbb{I}_N \setminus \mathbb{J}} \delta_{j \sigma(j)}$. Then (3.69) becomes

$$\sum_{\mathbb{J}\subset\mathbb{I}_N, \sharp\mathbb{J}=N'} \sum_{\sigma\in\mathscr{S}(\mathbb{J})} \mathrm{sgn}(\sigma)\mathrm{E}^{\mathbf{x}}\left[F_{N'}(\mathbf{B}_{\mathbb{J}}(t))\prod_{i\in\mathbb{J}}\mathscr{M}_{\xi}^{x_{\sigma(i)}}(T, B_i(T))\right]$$
$$= \sum_{\mathbb{J}\subset\mathbb{I}_N, \sharp\mathbb{J}=N'} \mathrm{E}^{\mathbf{x}}\left[F_{N'}(\mathbf{B}_{\mathbb{J}}(t)) \det_{i,j\in\mathbb{J}}[\mathscr{M}_{\xi}^{x_j}(T, B_i(T))]\right]$$
$$= \int_{\mathbb{W}^{\mathrm{A}}_{N'}} \xi^{\otimes N'}(d\mathbf{v})\mathrm{E}^{\mathbf{v}}\left[F_{N'}(\mathbf{B}_{N'}(t)) \det_{i,j\in\mathbb{I}_{N'}}[\mathscr{M}_{\xi}^{x_j}(T, B_i(T))]\right],$$

where equivalence in probability law of $(B_i(t))_{t\geq 0}$, $1 \leq i \leq N$ is used. This is the RHS of (3.68) and the proof is completed. □

In order to show applications of DMR, we will derive the density function at a single time and the two-time correlation function below. Let $\mathrm{C}_{\mathrm{c}}(\mathbb{R})$ be the set of all continuous real-valued functions with compact supports.

3.5.1 Density Function $\rho_{\xi}(t, x)$

The *density function* at a single time for $(\Xi, \mathbb{P}^{\xi})$, $\xi \in \mathfrak{M}_0$ is denoted by $\rho_{\xi}(t, x)$. It is defined as a continuous function of $x \in \mathbb{R}$ for $0 \leq t \leq T < \infty$ such that for any *test function*, $\chi \in \mathrm{C}_{\mathrm{c}}(\mathbb{R})$,

$$\mathbb{E}^{\xi}\left[\int_{\mathbb{R}} \chi(x)\Xi(t, dx)\right] = \int_{\mathbb{R}} dx\, \chi(x)\rho_{\xi}(t, x). \tag{3.70}$$

The test function χ is symmetrized as

$$g(\mathbf{x}) = \sum_{i=1}^{N} \chi(x_i),$$

which is applied as F to the DMR (3.63), and we obtain the equality

$$\mathbb{E}^{\xi}\left[\sum_{i=1}^{N}\chi(X_i(t))\right] = \mathrm{E}^{\mathbf{x}}\left[\sum_{i=1}^{N}\chi(B_i(t))\mathscr{D}_{\xi}(T, \mathbf{B}(T))\right], \quad 0 \leq t \leq T < \infty. \tag{3.71}$$

The LHS of (3.71) gives

$$\mathbb{E}^{\xi}\left[\sum_{i=1}^{N}\chi(X_i(t))\right] = \mathbb{E}^{\xi}\left[\int_{\mathbb{R}} \chi(x)\Xi(t, dx)\right]$$

by (3.52). On the other hand, the RHS of (3.71) is reduced by Lemma 3.7 as

$$\begin{aligned}\sum_{i=1}^{N} \mathrm{E}^{\mathbf{x}}[\chi(B_i(t))\mathscr{D}_\xi(T,\mathbf{B}(T))] &= \int_{\mathbb{R}} \xi(dv)\mathrm{E}^{v}[\chi(B(t))\mathscr{M}_\xi^{v}(t, B(t))] \\ &= \int_{\mathbb{R}} \xi(dv) \int_{\mathbb{R}} dx\, \chi(x) p(t, x|v)\mathscr{M}_\xi^{v}(t, x).\end{aligned}$$

By Fubini's theorem, we can rewrite it as $\int_{\mathbb{R}} dx\, \chi(x)\mathscr{G}_\xi(t, x; t, x)$, where

$$\mathscr{G}_\xi(s, x; t, y) = \int_{\mathbb{R}} \xi(dv) p(s, x|v)\mathscr{M}_\xi^{v}(t, y). \tag{3.72}$$

Then (3.70) gives

$$\rho_\xi(t, x) = \mathscr{G}_\xi(t, x; t, x), \quad x \in \mathbb{R}, \quad t \geq 0. \tag{3.73}$$

3.5.2 Two-Time Correlation Function $\rho_\xi(s, x; t, y)$

For $0 \leq t_1 < t_2 \leq T < \infty$, set

$$g_1(\mathbf{x}) = \sum_{i=1}^{N} \chi_1(x_i), \quad g_2(\mathbf{x}) = \sum_{i=1}^{N} \chi_2(x_i),$$

where $\chi_m \in \mathrm{C}_{\mathrm{c}}(\mathbb{R})$, $m = 1, 2$, and put

$$F(\Xi(\cdot)) = \prod_{m=1}^{2} g_m(\mathbf{X}(t_m)).$$

If we apply this to DMR, (3.63), we obtain the equality

$$\begin{aligned}&\mathbb{E}^{\xi}\left[\sum_{i=1}^{N}\sum_{j=1}^{N} \chi_1(X_i(t_1))\chi_2(X_j(t_2))\right] \\ &= \mathrm{E}^{\mathbf{x}}\left[\sum_{i=1}^{N}\sum_{j=1}^{N} \chi_1(B_i(t_1))\chi_2(B_j(t_2))\mathscr{D}_\xi(T,\mathbf{B}(T))\right], \quad 0 \leq t \leq T < \infty.\end{aligned} \tag{3.74}$$

The LHS of (3.74) defines the two-time correlation function $\rho_\xi(s, x; t, y)$ as

$$\mathbb{E}^\xi\left[\sum_{i=1}^N\sum_{j=1}^N \chi_1(X_i(t_1))\chi_2(X_j(t_2))\right] = \int_{\mathbb{R}^2} dx_1 dx_2\, \chi_1(x_1)\chi_2(x_2)\rho_\xi(t_1, x_1; t_2, x_2). \tag{3.75}$$

On the other hand, the RHS of (3.74) gives

$$\begin{aligned}
&\sum_{i=1}^N\sum_{j=1}^N \mathrm{E}^{\mathbf{x}}[\chi_1(B_i(t_1))\chi_2(B_j(t_2))\mathscr{D}_\xi(T, \mathbf{B}(T))] \\
&\quad = \sum_{\substack{1\le i,j\le N,\\ i\neq j}} \mathrm{E}^{\mathbf{x}}[\chi_1(B_i(t_1))\chi_2(B_j(t_2))\mathscr{D}_\xi(T, \mathbf{B}(T))] \\
&\quad + \sum_{1\le i\le N} \mathrm{E}^{\mathbf{x}}[\chi_1(B_i(t_1))\chi_2(B_i(t_2))\mathscr{D}_\xi(T, \mathbf{B}(T))].
\end{aligned}$$

By the reducibility of DMR given by Lemma 3.7, the last expression becomes

$$\begin{aligned}
&\int_{\mathbb{R}^2} \xi^{\otimes 2}(d\mathbf{v}) \\
&\quad \times \mathrm{E}^{(v_1,v_2)}\left[\chi_1(B_1(t_1))\chi_2(B_2(t_2)) \det\begin{pmatrix} \mathscr{M}_\xi^{v_1}(T, B_1(T)) & \mathscr{M}_\xi^{v_1}(T, B_2(T)) \\ \mathscr{M}_\xi^{v_2}(T, B_1(T)) & \mathscr{M}_\xi^{v_2}(T, B_2(T)) \end{pmatrix}\right] \\
&+ \int_{\mathbb{R}} \xi(dv)\mathrm{E}^v[\chi_1(B(t_1))\chi_2(B(t_2))\mathscr{M}_\xi^v(T, B(T))]
\end{aligned}$$

If we use the martingale property (i) of $\mathscr{M}_\xi^v$ in Proposition 3.2, it is written as

$$\begin{aligned}
&\int_{\mathbb{R}^2} \xi^{\otimes 2}(d\mathbf{v}) \\
&\quad \times \mathrm{E}^{(v_1,v_2)}\left[\chi_1(B_1(t_1))\chi_2(B_2(t_2)) \det\begin{pmatrix} \mathscr{M}_\xi^{v_1}(t_1, B_1(t_1)) & \mathscr{M}_\xi^{v_1}(t_2, B_2(t_2)) \\ \mathscr{M}_\xi^{v_2}(t_1, B_1(t_1)) & \mathscr{M}_\xi^{v_2}(t_2, B_2(t_2)) \end{pmatrix}\right] \\
&+ \int_{\mathbb{R}} \xi(dv)\mathrm{E}^v[\chi_1(B(t_1))\chi_2(B(t_2))\mathscr{M}_\xi^v(t_2, B(t_2))].
\end{aligned}$$

By Fubini's theorem, this is equal to (see Exercise 3.10)

$$\int_{\mathbb{R}^2} dx_1 dx_2\, \chi_1(x_1)\chi_2(x_2) \det \begin{pmatrix} \mathscr{G}_\xi(t_1, x_1; t_1, x_1) & \mathscr{G}_\xi(t_1, x_1; t_2, x_2) \\ \mathscr{G}_\xi(t_2, x_2; t_1, x_1) & \mathscr{G}_\xi(t_2, x_2; t_2, x_2) \end{pmatrix}$$
$$+ \int_{\mathbb{R}^2} dx_1 dx_2\, \chi_1(x_1)\chi_2(x_2) \mathscr{G}_\xi(t_1, x_1; t_2, x_2) p(t_2 - t_1, x_2|x_1)$$
$$= \int_{\mathbb{R}^2} dx_1 dx_2\, \chi_1(x_1)\chi_2(x_2)$$
$$\times \det \begin{pmatrix} \mathscr{G}_\xi(t_1, x_1; t_1, x_1) & \mathscr{G}_\xi(t_1, x_1; t_2, x_2) \\ \mathscr{G}_\xi(t_2, x_2; t_1, x_1) - p(t_2 - t_1, x_2|x_1) & \mathscr{G}_\xi(t_2, x_2; t_2, x_2) \end{pmatrix}.$$

Since this is equal to (3.75), the two-time correlation function is determined as

$$\rho_\xi(s, x; t, y) = \det \begin{pmatrix} \mathbb{K}_\xi(s, x; s, x) & \mathbb{K}_\xi(s, x; t, y) \\ \mathbb{K}_\xi(t, y; s, x) & \mathbb{K}_\xi(t, y; t, y) \end{pmatrix} \tag{3.76}$$

for $0 \le s < t < \infty$, $x, y \in \mathbb{R}$, where

$$\mathbb{K}_\xi(s, x; t, y) = \mathscr{G}_\xi(s, x; t, y) - \mathbf{1}_{(s>t)} p(s - t, x|y). \tag{3.77}$$

3.6 Determinantal Process

In the previous section, the density function at a single time $\rho_\xi(t, x)$ and the two-time (and two-point) correlation function $\rho_\xi(s, x; t, y)$ were defined by (3.70) and (3.75), respectively. In order to give a general definition of *spatio-temporal correlation functions* here we consider the Laplace transformations of the multitime joint distribution functions of $(\Xi, \mathbb{P}^\xi)$. For any integer $M \in \mathbb{N}$, a sequence of times $\mathbf{t} = (t_1, \dots, t_M) \in [0, \infty)^M$ with $0 \le t_1 < \dots < t_M < \infty$, and a sequence of functions $\mathbf{f} = (f_{t_1}, \dots, f_{t_M}) \in \mathrm{C}_\mathrm{c}(\mathbb{R})^M$, let

$$\Psi_\xi^{\mathbf{t}}[\mathbf{f}] \equiv \mathbb{E}^\xi \left[\exp \left\{ \sum_{m=1}^{M} \int_{\mathbb{R}} f_{t_m}(x) \Xi(t_m, dx) \right\} \right]. \tag{3.78}$$

By (3.52), if we set test functions as

$$\chi_{t_m}(\cdot) = e^{f_{t_m}(\cdot)} - 1, \quad 1 \le m \le M, \tag{3.79}$$

we can rewrite (3.78) in the form

$$\Psi_\xi^{\mathbf{t}}[\mathbf{f}] = \mathbb{E}^\xi \left[\prod_{m=1}^{M} \prod_{i=1}^{N} \{1 + \chi_{t_m}(X_i(t_m))\} \right]. \tag{3.80}$$

We expand this with respect to test functions and define the spatio-temporal correlation functions $\{\rho_\xi\}$ as coefficients,

$$\Psi_\xi^{\mathbf{t}}[\mathbf{f}] = \sum_{\substack{0 \le N_m \le N, \\ 1 \le m \le M}} \int_{\prod_{m=1}^M \mathbb{W}_{N_m}^{\mathrm{A}}} \prod_{m=1}^M d\mathbf{x}_{N_m}^{(m)} \prod_{i=1}^{N_m} \chi_{t_m}\left(x_i^{(m)}\right) \rho_\xi\left(t_1, \mathbf{x}_{N_1}^{(1)}; \dots; t_M, \mathbf{x}_{N_M}^{(M)}\right), \tag{3.81}$$

where $\mathbf{x}_{N_m}^{(m)}$ denotes $(x_1^{(m)}, \dots, x_{N_m}^{(m)})$ and $d\mathbf{x}_{N_m}^{(m)} = \prod_{i=1}^{N_m} dx_i^{(m)}$, $1 \le m \le M$. Here the empty products equal 1 by convention and the term with $N_m = 0$, $1 \le \forall m \le M$ is considered to be 1. The previous two examples $\rho_\xi(t, x)$ and $\rho_\xi(s, x; t, y)$ are special cases in which we set $M = 1$, $t_1 = t$, $N_1 = 1$, $x_1^{(1)} = x$, and $M = 2$, $t_1 = s$, $t_2 = t$, $N_1 = N_2 = 1$, $x_1^{(1)} = x$, $x_1^{(2)} = y$, respectively. The function $\Psi_\xi^{\mathbf{t}}[\mathbf{f}]$ is a *generating function of correlation functions*.

Assume that the two processes $\Xi(t)$, $t \ge 0$ and $\widetilde{\Xi}(t)$, $t \ge 0$ are defined in the same probability space and both are started at the same configuration ξ. If they have the same generating function of correlations $\Psi_\xi^{\mathbf{t}}[\mathbf{f}]$ for any $M \in \mathbb{N}$, $\mathbf{t} \in [0, \infty)^M$ with $0 \le t_1 < \cdots < t_M < \infty$, and $\mathbf{f} \in \mathrm{C}_{\mathrm{c}}(\mathbb{R})^M$, then these two processes are specified by the same correlation functions. In this case, we say that the processes $\Xi(t)$, $t \ge 0$ and $\widetilde{\Xi}(t)$, $t \ge 0$ are *equivalent in the sense of finite-dimensional distributions*.

Given an integral kernel, $\mathbf{K}(s, x; t, y)$, $(s, x), (t, y) \in [0, \infty) \times \mathbb{R}$, and a sequence of functions $(\chi_{t_1}, \dots, \chi_{t_M}) \in \mathrm{C}_{\mathrm{c}}(\mathbb{R})^M$, $M \in \mathbb{N}$, the *Fredholm determinant* associated with $\mathbf{K}$ and $(\chi_{t_m})_{m=1}^M$ is defined as

$$\begin{aligned} &\underset{\substack{(s,t) \in \{t_1, \dots, t_M\}^2, \\ (x,y) \in \mathbb{R}^2}}{\mathrm{Det}} \Big[\delta_{st}\delta_x(\{y\}) + \mathbf{K}(s, x; t, y)\chi_t(y)\Big] \\ &= \sum_{\substack{0 \le N_m \le N, \\ 1 \le m \le M}} \int_{\prod_{m=1}^M \mathbb{W}_{N_m}^{\mathrm{A}}} \prod_{m=1}^M d\mathbf{x}_{N_m}^{(m)} \prod_{k=1}^{N_m} \chi_{t_m}\left(x_k^{(m)}\right) \underset{\substack{1 \le i \le N_m, 1 \le j \le N_n, \\ 1 \le m, n \le M}}{\det} \Big[\mathbf{K}(t_m, x_i^{(m)}; t_n, x_j^{(n)})\Big]. \end{aligned} \tag{3.82}$$

If we consider the simplest case where $M = 1$ and $t_1 = t \in [0, \infty)$ in (3.82), we have

$$\underset{(x,y) \in \mathbb{R}^2}{\mathrm{Det}} \Big[\delta_x(\{y\}) + \mathbf{K}(t, x; t, y)\chi_t(y)\Big] = \sum_{N'=0}^N \int_{\mathbb{W}_{N'}^{\mathrm{A}}} d\mathbf{x}_{N'} \prod_{k=1}^{N'} \chi_t(x_k) \det_{1 \le i, j \le N'} [\mathbf{K}(t, x_i; t, x_j)].$$

Given $\mathbf{v} = (v_1, \dots, v_N) \in \mathbb{W}_N^{\mathrm{A}}$, put $\chi_t(x) = \sum_{\ell=1}^N \widehat{\chi}_\ell \delta_{v_\ell}(x)$ with $\widehat{\chi}_\ell \in \mathbb{R}$, $1 \le \ell \le N$. In this case the above is equal to

$$\sum_{N'=0}^N \sum_{\mathbb{J} \subset \mathbb{I}_N, \sharp\mathbb{J} = N'} \prod_{k \in \mathbb{J}} \widehat{\chi}_k \det_{i, j \in \mathbb{J}} [K_{ij}]$$

with $K_{ij} = \mathbf{K}(t, v_i; t, v_j)$, $1 \le i, j \le N$. This is obtained as the *Fredholm expansion formula* of $\det_{1 \le i,j \le N}[\delta_{ij} + K_{ij}\widehat{\chi}_j]$ (Exercise 3.11). For this reason, (3.82) is called the Fredholm determinant. See, for instance, Chap. 21 in [101], Chap. 9 in [45], and Chap. 3 in [5] for more details of Fredholm determinants.

Definition 3.1 If any moment generating function (3.78) is given by a Fredholm determinant, the process $(\Xi, \mathbb{P}^\xi)$ is said to be *determinantal*. In this case all spatio-temporal correlation functions are given by determinants as

$$\rho_\xi\left(t_1, \mathbf{x}^{(1)}_{N_1}; \dots; t_M, \mathbf{x}^{(M)}_{N_M}\right) = \det_{\substack{1 \le i \le N_m, 1 \le j \le N_n, \\ 1 \le m,n \le M}} \left[\mathbb{K}_\xi(t_m, x_i^{(m)}; t_n, x_j^{(n)})\right], \tag{3.83}$$

$0 \le t_1 < \cdots < t_M < \infty$, $1 \le N_m \le N$, $\mathbf{x}^{(m)}_{N_m} \in \mathbb{R}^{N_m}$, $1 \le m \le M \in \mathbb{N}$. Here the integral kernel, $\mathbb{K}_\xi : ([0, \infty) \times \mathbb{R})^2 \mapsto \mathbb{R}$, is a function of the initial configuration ξ and is called the *correlation kernel*.

Remark 3.2 If the process $(\Xi, \mathbb{P}^\xi)$ is determinantal, then, for each specified time $0 \le t < \infty$, all spatial correlation functions are given by determinants as

$$\rho_\xi(\mathbf{x}_{N'}) = \det_{1 \le i,j \le N'} [\mathrm{K}(x_i, x_j)], \quad 1 \le N' \le N, \tag{3.84}$$

with $\mathrm{K}(x, y) = \mathbb{K}_\xi(t, x; t, y)$. In general a random integer-valued Radon measure in $\mathfrak{M}$ (resp. $\mathfrak{M}_0$) is called a *point process* (resp. *simple point process*). A simple point process is said to be a *determinantal point process* (or *Fermion point process*) with kernel K, if its spatial correlation functions exist and are given in the form (3.84). When K is symmetric, i.e., $\mathrm{K}(x, y) = \mathrm{K}(y, x)$, $x, y \in \mathbb{R}$, Soshnikov [126] and Shirai and Takahashi [123] gave sufficient conditions for K to be a correlation kernel of a determinantal point process (see also [3, 5, 57]). The notion of determinantal process given by Definition 3.1 is a dynamical extension of the determinantal point process [20, 82].

Here we give a lemma concerning the relevant part of the correlation kernel of a determinantal process.

Lemma 3.8 *Let Ξ and $\widetilde{\Xi}$ be determinantal processes with correlation kernels $\mathbb{K}$ and $\widetilde{\mathbb{K}}$, respectively. If there is a function $G(s, x)$, which is continuous with respect to $x \in \mathbb{R}$ for any fixed $s \in [0, \infty)$, such that*

$$\mathbb{K}(s, x; t, y) = \frac{G(s, x)}{G(t, y)} \widetilde{\mathbb{K}}(s, x; t, y), \quad (s, x), (t, y) \in [0, \infty) \times \mathbb{R}, \tag{3.85}$$

then Ξ and $\widetilde{\Xi}$ are equivalent in the sense of finite-dimensional distributions. In other words, any rational factor of the form given in (3.85) for correlation kernels is irrelevant, when we discuss correlation functions for determinantal processes.

Proof In general, the determinant of an $N \times N$ matrix $\mathrm{M} = (m_{ij})_{1 \le i,j \le N}$, $N \in \mathbb{N}$, is defined by $\sum_{\sigma \in \mathscr{S}_N} \mathrm{sgn}(\sigma) \prod_{i=1}^{N} m_{i\sigma(i)}$. Any permutation σ consists of exclusive cycles. If we write each cyclic permutation as

$$\mathsf{c} = \begin{pmatrix} a & b & \cdots & \omega \\ b & c & \cdots & a \end{pmatrix}$$

and the number of cyclic permutations in a given σ as $\ell(\sigma)$, then $\mathrm{sgn}(\sigma) = (-1)^{N-\ell(\sigma)}$ and the determinant of M is expressed as [101]

$$\det \mathrm{M} = \sum_{\sigma \in \mathscr{S}_N} (-1)^{N-\ell(\sigma)} \prod_{\mathsf{c}_i : 1 \le i \le \ell(\sigma)} \Big(m_{ab} m_{bc} \dots m_{\omega a} \Big).$$

It implies that, with given $a_1, a_2, \dots, a_N$, even if each element m_{ij} of the matrix M is replaced by $m_{ij} \times (a_i/a_j)$, the value of determinant is not changed. Then by Definition 3.1 of determinantal process, the present lemma is readily concluded. □

By Proposition 3.2, (3.80) has the DMR

$$\Psi_\xi^{\mathbf{t}}[\mathbf{f}] = \mathrm{E}^{\mathbf{x}} \left[\prod_{m=1}^{N} \prod_{i=1}^{N} \{1 + \chi_{t_m}(B_i(t_m))\} \mathscr{D}_\xi(T, \mathbf{B}(T)) \right], \tag{3.86}$$

where $T \ge t_M$ and $\xi = \sum_{i=1}^{N} \delta_{x_i}$. The following equality is established.

Lemma 3.9 *Let* $\mathbf{x} \in \mathbb{W}_N^{\mathrm{A}}$ *and* $\xi = \sum_{i=1}^{N} \delta_{x_i}$. *Then for any* $M \in \mathbb{N}$, $0 \le t_1 < \cdots < t_M \le T < \infty$, $\chi_{t_m} \in \mathrm{C}_c(\mathbb{R})$, $1 \le m \le M$, *the equality*

$$\begin{aligned} &\mathrm{E}^{\mathbf{x}} \left[\prod_{m=1}^{M} \prod_{i=1}^{N} \{1 + \chi_{t_m}(B_i(t_m))\} \mathscr{D}_\xi(T, \mathbf{B}(T)) \right] \\ &= \underset{\substack{(s,t) \in \{t_1, \dots, t_M\}^2, \\ (x,y) \in \mathbb{R}^2}}{\mathrm{Det}} \Big[\delta_{st} \delta_x(\{y\}) + \mathbb{K}_\xi(s, x; t, y) \chi_t(y) \Big] \end{aligned} \tag{3.87}$$

holds, where $\mathbb{K}_\xi$ *is given by (3.77) with (3.72).*

Proof The LHS of (3.87) is an expectation of a usual determinant $\mathscr{D}_\xi$ multiplied by test functions, while the RHS is a Fredholm determinant. Note that the expectation in the LHS will be calculated by performing integrals using the transition probability density p of BMs, $(B_i(t))_{t \ge 0}$, $1 \le i \le N$, as an integral kernel, while p is involved in the integral representation (3.72) for the correlation kernel $\mathbb{K}_\xi$ for the Fredholm determinant in the RHS. Therefore, quite simply, this equality is just obtained by applying Fubini's theorem and changing the order of integrals. Since the quantities in (3.87) are multivariate and the multitime joint distribution is considered, however, we also need combinatorial arguments to prove the equality. Since the full proof was given in [73], here we omit it. □

Now we arrive at the following characterization of the Dyson model $(\Xi, \mathbb{P}^{\xi})$.

Theorem 3.4 *For any finite and fixed initial configuration without multiple points, that is, for* $\xi \in \mathfrak{M}_0$, $\xi(\mathbb{R}) = N \in \mathbb{N}$, *the Dyson model is determinantal. Its correlation kernel is given by*

$$\mathbb{K}_{\xi}(s, x; t, y) = \mathscr{G}_{\xi}(s, x; t, y) - \mathbf{1}_{(s>t)} p(s - t, x|y), \quad (s, x), (t, y) \in [0, \infty) \times \mathbb{R} \tag{3.88}$$

with

$$\mathscr{G}_{\xi}(s, x; t, y) = \int_{\mathbb{R}} \xi(dv) p(s, x|v) \mathscr{M}_{\xi}^{v}(t, y). \tag{3.89}$$

Remark 3.3 We note that, for each chosen series of times $\mathbf{t} = (t_1, \ldots, t_M)$, $M \in \mathbb{N}$, $\Xi_{\mathbf{t}} \equiv \sum_{t \in \mathbf{t}} \delta_t \otimes \Xi(t)$ can be regarded as a determinantal point process on the spatio-temporal field $\mathbf{t} \times \mathbb{R}$ with the kernel $\mathbb{K}_{\xi}(s, x; t, y)$, $(s, x), (t, y) \in \mathbf{t} \times \mathbb{R}$. (It is called a determinantal M-component system in Sect. 5.9 of [45].) It should be remarked that the present correlation kernels (3.88) with (3.89) are asymmetric, $\mathbb{K}_{\xi}(s, x; t, y) \neq \mathbb{K}_{\xi}(t, y; s, x)$ due to the second terms $-\mathbf{1}_{(s>t)} p(s - t, x|y)$. Such form of asymmetric correlation kernels is said to be of the *Eynard–Mehta type* [20, 40, 84]. From the viewpoint of statistical physics, such asymmetry is necessary to describe *nonequilibrium systems* developing in time [59, 65, 114, 118]. General mathematical theory to give sufficient conditions for an asymmetric integral kernel to be a determinantal correlation kernel on $\mathbf{t} \times \mathbb{R}$ is, however, not yet known.

Derivations of the Eynard–Mehta type correlation kernels via DMRs were reported in [73] for the noncolliding squared Bessel process and the trigonometric extension of the Dyson model, in [75] for the elliptic extension of the Dyson model of type A, and in [39, 74] for the discrete-time and continuous-time systems of noncolliding random walks (the vicious walker models).

3.7 Constant-Drift Transformation of Dyson Model

Let b a real constant. We consider the Dyson model with constant drift (*drifted Dyson model*), which is given by the following system of SDEs,

$$dX_i(t) = dB_i^{x_i}(t) + b\,dt + \sum_{\substack{1 \le j \le N, \\ j \neq i}} \frac{dt}{X_i(t) - X_j(t)}, \quad t \ge 0, \quad 1 \le i \le N. \tag{3.90}$$

Let

$$h_N^{(b)}(t, \mathbf{x}) = \prod_{i=1}^{N} e^{bx_i - b^2 t/2} h_N(\mathbf{x}), \quad \mathbf{x} \in \mathbb{W}_N^{\mathrm{A}}, \quad t \ge 0, \tag{3.91}$$

where $h_N(\mathbf{x})$ is given by (3.31). Then we can prove that the transition probability density of the drifted Dyson model is given by

$$p_N^{(b)}(t-s,\mathbf{y}|\mathbf{x}) = \frac{h_N^{(b)}(t,\mathbf{y})}{h_N^{(b)}(s,\mathbf{x})} q_N(t-s,\mathbf{y}|\mathbf{x}), \quad \mathbf{x},\mathbf{y} \in \mathbb{W}_N^{\mathrm{A}}, \quad 0 \le s \le t, \tag{3.92}$$

where q_N is the Karlin–McGregor determinant given by (3.27) (Exercise 3.12).

By the multilinearity of determinants, we see that

$$\begin{aligned} h_N^{(b)}(t,\mathbf{y}) &= \det_{1\le i,j\le N} [m_{j-1}^{(b)}(t,y_i)] \\ &= \det_{1\le i,j\le N} \Big[\mathscr{I}[e^{bW} W^{j-1}|(t,y_i)] \Big], \end{aligned}$$

where (1.35) was used. Therefore, the linearity of the integral transformation $\mathscr{I}$ and Lemma 3.5 give

$$\begin{aligned} \frac{h_N^{(b)}(t,\mathbf{y})}{h_N^{(b)}(0,\mathbf{x})} &= \mathscr{I}\left[\prod_{k=1}^N e^{b(W_k-x_k)} \frac{h_N(\mathbf{W})}{h_N(\mathbf{x})} \bigg| (t,\mathbf{y}) \right] \\ &= \mathscr{I}\left[\prod_{k=1}^N e^{b(W_k-x_k)} \det_{1\le i,j\le N} [\Phi_\xi^{x_i}(W_j)] \bigg| (t,\mathbf{y}) \right] \\ &= \det_{1\le i,j\le N} [\mathscr{M}_\xi^{x_i,b}(t,y_j)], \end{aligned}$$

with $\xi = \sum_{i=1}^N \delta_{x_i}$ and

$$\mathscr{M}_\xi^{x,(b)}(t,y) = \mathscr{I}[e^{b(W-x)}\Phi_\xi^x(W)|(t,y)], \quad x,y \in \mathbb{R}, \quad t \ge 0. \tag{3.93}$$

The above results imply the following.

Let $(\Xi^{(b)}, \mathbb{P}^{\xi,(b)})$ be the Dyson model with constant drift (3.90), where the expectation with respect to $\mathbb{P}^{\xi,(b)}$ is written by $\mathbb{E}^{\xi,(b)}$ and the filtration generated by $\{\Xi^{(b)}(s) : s \le t\}$ is denoted by $(\mathscr{F}_{\Xi^{(b)}})_t$, $t \ge 0$.

Proposition 3.3 *The Dyson model with constant drift (3.90) has the following DMR: for any $(\mathscr{F}_{\Xi^{(b)}})_t$-measurable bounded function F, $0 \le t \le T < \infty$,*

$$\mathbb{E}^{\xi,(b)}[F(\Xi^{(b)}(\cdot)] = \mathrm{E}^{\mathbf{x}}\left[F\left(\sum_{i=1}^N \delta_{B_i(\cdot)} \right) \mathscr{D}_\xi^{(b)}(T,\mathbf{B}(T)) \right], \tag{3.94}$$

where $\xi = \sum_{i=1}^N \delta_{x_i}$ and

$$\mathscr{D}_\xi^{(b)}(t,\mathbf{y}) = \det_{1\le i,j\le N} [\mathscr{M}_\xi^{y_j,(b)}(t,y_i)], \quad \mathbf{y} \in \mathbb{W}_N^{\mathrm{A}}, \quad t \ge 0 \tag{3.95}$$

with (3.93).

Following the argument given in Sect. 3.6, we can conclude the following.

Proposition 3.4 *For any* $\xi \in \mathfrak{M}_0$, $\xi(\mathbb{R}) = N \in \mathbb{N}$, *the drifted Dyson model with a constant drift coefficient b, (3.90), is determinantal. Its correlation kernel is given by*

$$\mathbb{K}_\xi^{(b)}(s, x; t, y) = \mathscr{G}_\xi^{(b)}(s, x; t, y) - \mathbf{1}_{(s>t)} p(s-t, x|y), \quad (s, x), (t, y) \in [0, \infty) \times \mathbb{R}, \tag{3.96}$$

with

$$\mathscr{G}_\xi^{(b)}(s, x; t, y) = \int_{\mathbb{R}} \xi(dv)\, p(s, x|v) \mathscr{M}_\xi^{v,(b)}(t, y). \tag{3.97}$$

We note that (3.93) is related to the martingale function of the Dyson model without drift $\mathscr{M}_\xi^x$ as (Exercise 3.13)

$$\mathscr{M}_\xi^{x,(b)}(t, y) = e^{b(y-x)-b^2t/2} \mathscr{M}_\xi^x(t, y - bt), \quad x, y \in \mathbb{R}, \quad t \geq 0. \tag{3.98}$$

Then (3.96) is rewritten as

$$\mathbb{K}_\xi^{(b)}(s, x; t, y) = \frac{e^{-bx+b^2s/2}}{e^{-by+b^2t/2}} \widetilde{\mathbb{K}}_\xi^{(b)}(s, x; t, y)$$

with

$$\widetilde{\mathbb{K}}_\xi^{(b)}(s, x; t, y) = \int_{\mathbb{R}} \xi(dv)\, p^{(b)}(s, x|v) \mathscr{M}_\xi^v(t, y - bt) - \mathbf{1}_{(s>t)} p^{(b)}(s - t, x|y), \tag{3.99}$$

where $p^{(b)}$ is the drift transformation (1.29) of p. By Lemma 3.8, the correlation kernel (3.96) in Proposition 3.4 can be replaced by (3.99). For $p^{(b)}(s, x|v) = p(s, x - bs|v)$ and $p^{(b)}(s - t, x|y) = p(s - t, x - bs|y - bt)$, we arrive at the following result.

Theorem 3.5 *The correlation kernel of the drifted Dyson model with a constant drift coefficient b is given by*

$$\mathbb{K}_\xi(s, x - bs; t, y - bt), \quad (s, x), (t, y) \in [0, \infty) \times \mathbb{R}. \tag{3.100}$$

That is, the constant-drift transformation of the Dyson model with a drift coefficient b is obtained by performing the Galilean transformation with a constant velocity b, $(t, x) \to (t, x - bt)$, *in the spatio-temporal coordinates of the Dyson model without drift.*

The case where the drift coefficients of N particles depend on individual particles and are given as $\mathbf{b} = (b_1, \dots, b_N) \in \mathbb{W}_N^{\mathrm{A}}$ is discussed in [14, 71]. See [130, 131] for the connection with the *biorthogonal ensembles* of random matrix theory [18, 102] and the *Chern–Simons theory* [31, 96, 97, 134].

3.8 Generalization for Initial Configuration with Multiple Points

For general $\xi = \sum_{i=1}^N \delta_{x_i} \in \mathfrak{M}$ with $\xi(\mathbb{R}) = N < \infty$, define $\mathrm{supp}\, \xi = \{x \in \mathbb{R} : \xi(x) > 0\}$ and let $\xi_*(\cdot) = \sum_{v \in \mathrm{supp}\, \xi} \delta_v(\cdot)$. For $s \in [0, \infty)$, $v, x \in \mathbb{R}$, $z, \zeta \in \mathbb{C}$, let

$$\phi_\xi^v((s,x); z, \zeta) = \frac{p(s,x|\zeta)}{p(s,x|v)} \frac{1}{z-\zeta} \prod_{i=1}^N \frac{z-x_i}{\zeta - x_i}, \tag{3.101}$$

and

$$\begin{aligned}\Phi_\xi^v((s,x); z) &= \frac{1}{2\pi\sqrt{-1}} \oint_{C(\delta_v)} d\zeta\, \phi_\xi^v((s,x); z, \zeta) \\ &= \mathrm{Res}\left[\phi_\xi^v((s,x); z, \zeta); \zeta = v\right],\end{aligned} \tag{3.102}$$

where $C(\delta_v)$ is a closed contour on the complex plane $\mathbb{C}$ encircling the point v once in the positive direction. This function (3.102) is entire with respect to $z \in \mathbb{C}$ parameterized by $(s, x) \in [0, \infty) \times \mathbb{R}$ in addition to $v \in \mathbb{C}, \xi \in \mathfrak{M}$. Remark that the polynomial function $\Phi_\xi^u(z)$ defined by (3.60) is parameterized only by $u \in \mathbb{C}$ and $\xi \in \mathfrak{M}_0$. In the paper [84], this entire function was constructed by combining the multiple Hermite polynomials of type I and type II [60] using their integral representations [17]. Here we start from this entire function and consider its $\mathscr{I}$-transformation,

$$\mathscr{M}_\xi^v((s,x)|(t,y)) = \mathscr{I}\left[\Phi_\xi^v((s,x); W)\Big|(t,y)\right], \quad (s,x), (t,y) \in [0,\infty) \times \mathbb{R}, \tag{3.103}$$

which provides a martingale, if we put $y = B(t),\ t \geq 0$. Then it is easy to see that (3.88) with (3.89) is rewritten as

$$\mathbb{K}_\xi(s,x;t,y) = \int_{\mathbb{R}} \xi_*(dv) p(s,x|v) \mathscr{M}_\xi^v((s,x)|(t,y)) - \mathbf{1}(s>t) p(s-t, x|y), \tag{3.104}$$

$(s, x), (t, y) \in [0, \infty) \times \mathbb{R}$ (Exercise 3.14).

We note that the kernel (3.104) with (3.103) is bounded and integrable also for $\xi \in \mathfrak{M} \backslash \mathfrak{M}_0$. Therefore, the spatio-temporal correlations are given by (3.83) for any $0 \leq t_1 < \cdots < t_M < \infty$, $M \in \mathbb{N}$ and the finite-dimensional distributions are determined.

Proposition 3.5 *Also for $\xi \in \mathfrak{M} \backslash \mathfrak{M}_0$, the determinantal processes with the correlation kernels (3.104) are well-defined.*

The complete proof of this proposition was given in Sect. 4.1 of [84]. The above extension will provide the *entrance laws* for the processes $(\Xi(t), t > 0, \mathbb{P}^{\xi})$ in the sense of Sect. 12.4 in [116].

In order to give an example of Proposition 3.5, here we study the extreme case where all N points are concentrated on an origin,

$$\xi = N\delta_0 \iff \xi_* = \delta_0 \quad \text{with} \quad \xi(\{0\}) = N. \tag{3.105}$$

For (3.105), (3.101) and (3.102) become

$$\begin{aligned} \phi^0_{N\delta_0}((s,x); z, \zeta) &= \frac{p(s,x|\zeta)}{p(s,x|0)} \frac{1}{z-\zeta} \left(\frac{z}{\zeta}\right)^N \\ &= \frac{p(s,x|\zeta)}{p(s,x|0)} \sum_{\ell=0}^{\infty} \frac{z^{N-\ell-1}}{\zeta^{N-\ell}}, \end{aligned}$$

and

$$\begin{aligned} \Phi^0_{N\delta_0}((s,x); z) &= \frac{1}{p(s,x|0)} \sum_{\ell=0}^{\infty} z^{N-\ell-1} \frac{1}{2\pi\sqrt{-1}} \oint_{C(\delta_0)} d\zeta \, \frac{p(s,x|\zeta)}{\zeta^{N-\ell}} \\ &= \frac{1}{p(s,x|0)} \sum_{\ell=0}^{N-1} z^{N-\ell-1} \frac{1}{2\pi\sqrt{-1}} \oint_{C(\delta_0)} d\zeta \, \frac{p(s,x|\zeta)}{\zeta^{N-\ell}}, \end{aligned} \tag{3.106}$$

since the integrands are holomorphic when $\ell \geq N$.

For BM with the transition probability density (1.1), (3.106) gives

$$\begin{aligned} \Phi^0_{N\delta_0}((s,x); z) &= \sum_{\ell=0}^{N-1} z^{N-\ell-1} \frac{1}{2\pi\sqrt{-1}} \oint_{C(\delta_0)} d\zeta \, \frac{e^{x\zeta/s-\zeta^2/2s}}{\zeta^{N-\ell}} \\ &= \sum_{\ell=0}^{N-1} \left(\frac{z}{\sqrt{2s}}\right)^{N-\ell-1} \frac{1}{2\pi\sqrt{-1}} \oint_{C(\delta_0)} d\eta \, \frac{e^{2(x/\sqrt{2s})\eta-\eta^2}}{\eta^{N-\ell}} \\ &= \sum_{\ell=0}^{N-1} \left(\frac{z}{\sqrt{2s}}\right)^{N-\ell-1} \frac{1}{(N-\ell-1)!} H_{N-\ell-1}\left(\frac{x}{\sqrt{2s}}\right), \end{aligned}$$

where we have used the contour integral representation of the Hermite polynomials (1.118). Thus its integral transformation is calculated as

$$\begin{aligned}
&\mathscr{I}\left[\Phi^0_{N\delta_0}((s,x);W)\big|(t,y)\right] \\
&= \sum_{\ell=0}^{N-1} \frac{1}{(N-\ell-1)!} H_{N-\ell-1}\left(\frac{x}{\sqrt{2s}}\right) \frac{1}{(2s)^{(N-\ell-1)/2}} \mathscr{I}[W^{N-\ell-1}|(t,y)] \\
&= \sum_{\ell=0}^{N-1} \frac{1}{(N-\ell-1)!} H_{N-\ell-1}\left(\frac{x}{\sqrt{2s}}\right) \frac{1}{(2s)^{(N-\ell-1)/2}} m_{N-\ell-1}(t,y) \\
&= \sum_{\ell=0}^{N-1} \frac{1}{(N-\ell-1)!2^{N-\ell-1}} \left(\frac{t}{s}\right)^{(N-\ell-1)/2} H_{N-\ell-1}\left(\frac{x}{\sqrt{2s}}\right) H_{N-\ell-1}\left(\frac{y}{\sqrt{2t}}\right),
\end{aligned}$$

where we have used Lemma 1.1 and (1.26) in Sect. 1.2. Then we obtain the following,

$$\begin{aligned}
\mathscr{M}^0_{N\delta_0}((s,x)|(t,B(t))) &= \sum_{n=0}^{N-1} \frac{1}{n!2^n} m_n(s,x) m_n(t,B(t)) \\
&= \sqrt{\pi} e^{x^2/4s + B(t)^2/4t} \sum_{n=0}^{N-1} \left(\frac{t}{s}\right)^{n/2} \varphi_n\left(\frac{x}{\sqrt{2s}}\right) \varphi_n\left(\frac{B(t)}{\sqrt{2t}}\right), \qquad (3.107)
\end{aligned}$$

where

$$\varphi_n(x) = \frac{1}{\sqrt{\sqrt{\pi}2^n n!}} H_n(x) e^{-x^2/2}, \quad x \in \mathbb{R}, \quad n \in \mathbb{N}_0, \qquad (3.108)$$

are the *Hermite orthonormal functions* on $\mathbb{R}$,

$$\int_{\mathbb{R}} dx\, \varphi_n(x)\varphi_m(x) = \delta_{nm}, \quad n, m \in \mathbb{N}_0. \qquad (3.109)$$

(See (1.116).) The following expression for the transition probability density (1.1) of BM is known as *Mehler's formula* (Exercise 3.15), for $s \geq t$,

$$p(s-t, x|y) = \frac{e^{-x^2/4s}}{e^{-y^2/4t}} \frac{1}{\sqrt{2s}} \sum_{n=0}^{\infty} \left(\frac{t}{s}\right)^{n/2} \varphi_n\left(\frac{x}{\sqrt{2s}}\right) \varphi_n\left(\frac{y}{\sqrt{2t}}\right). \qquad (3.110)$$

Since $m_n, n \in \mathbb{N}_0$ are the fundamental martingale polynomials associated with BM, the process (3.107) is a continuous martingale. Then we see that

$$\mathrm{E}\left[\mathscr{M}^0_{N\delta_0}((s,x)|(t,B(t)))\right] = \mathrm{E}\left[\mathscr{M}^0_{N\delta_0}((s,x)|(0,B(0)))\right] = 1$$

for $(s,x) \in [0,\infty) \times \mathbb{R}$, $0 \leq t < \infty$.

By the formula (3.104), we obtain the correlation kernel as

$$\mathbb{K}_{N\delta_0}(s,x;t,y) = p(s,x|0)\mathscr{M}^0_{N\delta_0}((s,x)|(t,y)) - \mathbf{1}(s>t)p(s-t,x|y)$$
$$= \frac{e^{-x^2/4s}}{e^{-y^2/4t}}\mathbf{K}^{(N)}_{\text{Hermite}}(s,x;t,y) \qquad (3.111)$$

with

$$\mathbf{K}^{(N)}_{\text{Hermite}}(s,x;t,y) = \frac{1}{\sqrt{2s}}\sum_{n=0}^{N-1}\left(\frac{t}{s}\right)^{n/2}\varphi_n\left(\frac{x}{\sqrt{2s}}\right)\varphi_n\left(\frac{y}{\sqrt{2t}}\right)$$
$$-\mathbf{1}(s>t)\frac{1}{\sqrt{2s}}\sum_{n=0}^{\infty}\left(\frac{t}{s}\right)^{n/2}\varphi_n\left(\frac{x}{\sqrt{2s}}\right)\varphi_n\left(\frac{y}{\sqrt{2t}}\right).$$
$$= \begin{cases} \dfrac{1}{\sqrt{2s}}\displaystyle\sum_{n=0}^{N-1}\left(\frac{t}{s}\right)^{n/2}\varphi_n\left(\frac{x}{\sqrt{2s}}\right)\varphi_n\left(\frac{y}{\sqrt{2t}}\right) & \text{for } s \le t, \\ -\dfrac{1}{\sqrt{2s}}\displaystyle\sum_{n=N}^{\infty}\left(\frac{t}{s}\right)^{n/2}\varphi_n\left(\frac{x}{\sqrt{2s}}\right)\varphi_n\left(\frac{y}{\sqrt{2t}}\right) & \text{for } s > t, \end{cases} \qquad (3.112)$$

where Mehler's formula (3.110) was used. By Lemma 3.8, the factor $e^{-x^2/4s}/e^{-y^2/4t}$ in (3.111) is irrelevant for determinantal processes. The kernel $\mathbf{K}^{(N)}_{\text{Hermite}}$ is known as the *extended Hermite kernel* (see, for instance, Exercise 11.6.3 in [45]).

The equal-time correlation kernel

$$\mathrm{K}^{(N,t)}_{\text{Hermite}}(x,y) \equiv \mathbf{K}^{(N)}_{\text{Hermite}}(t,x;t,y)$$
$$= \frac{1}{\sqrt{2t}}\sum_{n=0}^{N-1}\varphi_n\left(\frac{x}{\sqrt{2t}}\right)\varphi_n\left(\frac{y}{\sqrt{2t}}\right)$$

has the following expression (see Exercises 3.16 and 3.17),

$$\mathrm{K}^{(N,t)}_{\text{Hermite}}(x,y) = \sqrt{\frac{N}{2}}\frac{\varphi_N(x/\sqrt{2t})\varphi_{N-1}(y/\sqrt{2t}) - \varphi_{N-1}(x/\sqrt{2t})\varphi_N(y/\sqrt{2t})}{x-y},$$
$$\text{if } x \neq y, \qquad (3.113)$$

$$\mathrm{K}^{(N,t)}_{\text{Hermite}}(x,x)$$
$$= \frac{1}{\sqrt{2t}}\left[N\left\{\varphi_N\left(\frac{x}{\sqrt{2t}}\right)\right\}^2 - \sqrt{N(N+1)}\varphi_{N-1}\left(\frac{x}{\sqrt{2t}}\right)\varphi_{N+1}\left(\frac{x}{\sqrt{2t}}\right)\right]. \qquad (3.114)$$

This spatial correlation kernel is a special case of the *Christoffel–Dorboux kernel* (see, for instance, Chap. 9 in [45] and Chap. 3 in [5]). It is called the *Hermite kernel* and defines the determinantal point process [123, 124, 126] on $\mathbb{R}$ such that a spatial correlation function is given by

$$\rho_{\text{Hermite}}^{(N,t)}(\mathbf{x}_{N'}) = \det_{1 \le i,j \le N'} \left[\mathrm{K}_{\text{Hermite}}^{(N,t)}(x_i, x_j) \right] \tag{3.115}$$

for any $1 \le N' \le N$ and $\mathbf{x}_{N'} = (x_1, \ldots, x_{N'}) \in \mathbb{R}^{N'}, t > 0$. We write the probability measure of this determinantal point process as $\mathrm{P}_{\text{Hermite}}^{(N,t)}$.

3.9 Wigner's Semicircle Law and Scaling Limits

In this section, we consider a determinantal process with the extended Hermite kernel $\mathbf{K}_{\text{Hermite}}^{(N)}$ given by (3.112), where N is the number of particles. The correlation functions are denoted by $\{\rho_{\text{Hermite}}^{(N)}(\cdot)\}$.

3.9.1 Wigner's Semicircle Law

The density function at time $t \ge 0$ and position $x \in \mathbb{R}$ is given by

$$\begin{aligned}
\rho_{\text{Hermite}}^{(N)}(t, x) &= \mathrm{K}_{\text{Hermite}}^{(N,t)}(x, x) \\
&= \frac{1}{\sqrt{2t}} \sum_{n=0}^{N-1} \varphi_n \left(\frac{x}{\sqrt{2t}} \right)^2 \\
&= \frac{1}{\sqrt{2t}} \left[N \varphi_N \left(\frac{x}{\sqrt{2t}} \right)^2 - \sqrt{N(N+1)} \varphi_{N-1} \left(\frac{x}{\sqrt{2t}} \right) \varphi_{N+1} \left(\frac{x}{\sqrt{2t}} \right) \right].
\end{aligned} \tag{3.116}$$

It is easy to verify that

$$\int_{-\infty}^{\infty} \rho_{\text{Hermite}}^{(N)}(t, x) dx = N$$

by the orthonormality (3.109) of $\{\varphi_n(x)\}_{n \in \mathbb{N}_0}$. We will obtain estimations for the asymptotics at $N \to \infty$. The following formulas are derived from Theorem 8.22.9 (a) and (b) in Chap. 8 of [129]. Let ε and ω be fixed positive numbers. We have

(i) $$\begin{aligned}
\varphi_N \left(\sqrt{2N+1} \cos \phi \right) &= \frac{1}{\sqrt{\pi \sin \phi}} \left(\frac{2}{N} \right)^{1/4} \\
&\times \left\{ \sin \left[\left(\frac{N}{2} + \frac{1}{4} \right) (\sin 2\phi - 2\phi) + \frac{3}{4}\pi \right] + \mathscr{O}\left(\frac{1}{N} \right) \right\}, \quad \varepsilon \le \phi \le \pi - \varepsilon
\end{aligned}$$

$$\text{(ii)} \quad \varphi_N\left(\sqrt{2N+1}\cosh\phi\right) = \frac{1}{\sqrt{2\pi\sinh\phi}}\left(\frac{1}{2N}\right)^{1/4}$$
$$\times \exp\left[\left(\frac{N}{2}+\frac{1}{4}\right)(2\phi - \sinh 2\phi) + \frac{3}{4}\pi\right]\left\{1+\mathcal{O}\left(\frac{1}{N}\right)\right\}, \quad \varepsilon \le \phi \le \omega.$$

Using these expressions, we have the asymptotics of the density profile at $N \to \infty$,

$$\rho^{(N)}_{\text{Hermite}}(t,x) \simeq \begin{cases} \dfrac{1}{\pi\sqrt{2t}}\sqrt{2N - \dfrac{x^2}{2t}}, & \text{if } -2\sqrt{Nt} \le x \le 2\sqrt{Nt}, \\ 0, & \text{otherwise.} \end{cases} \tag{3.117}$$

The distribution of N particles has a finite support, whose interval $\propto \sqrt{N}$, and thus $\rho^{(N)}_{\text{Hermite}}(t,x) \sim \sqrt{N} \to \infty$ as $N \to \infty$ for fixed $0 < t < \infty$. If we set $x = 2\sqrt{Nt}\xi$, we see that

$$\lim_{N\to\infty} \frac{1}{N}\rho^{(N)}_{\text{Hermite}}(t, 2\sqrt{Nt}\xi)dx = \begin{cases} \dfrac{2}{\pi}\sqrt{1-\xi^2}\,d\xi, & \text{if } -1 \le \xi \le 1, \\ 0, & \text{otherwise,} \end{cases} \tag{3.118}$$

which is known as *Wigner's semicircle law* [101].

In the following, we consider the scaling limits, in which the long-term limit $t \to \infty$ is taken at the same time with $N \to \infty$.

3.9.2 Bulk Scaling Limit and Homogeneous Infinite System

First we consider the central region $x \simeq 0$ in the semicircle-shaped profile of particle density in the scaling limit

$$t \simeq \frac{N}{\pi^2} \to \infty. \tag{3.119}$$

In this limit the system becomes homogeneous also in space with a constant density $\rho = 1$. We call this the *bulk scaling limit.*

Proposition 3.6 *For any $M \in \mathbb{N}$, any sequence $\{N_m\}_{m=1}^M$ of positive integers, and any strictly increasing sequence $\{s_m\}_{m=1}^M$ of positive numbers,*

$$\lim_{N\to\infty} \rho^{(N)}_{\text{Hermite}}\left(\frac{N}{\pi^2}+2s_1, \mathbf{x}^{(1)}_{N_1}; \cdots ; \frac{N}{\pi^2}+2s_M, \mathbf{x}^{(M)}_{N_M}\right)$$
$$= \det_{\substack{1\le i\le N_m, 1\le j\le N_n,\\ 1\le m,n\le M}} \left[\mathbf{K}_{\sin}(s_m, x_i^{(m)}; s_n, x_j^{(n)})\right]$$
$$\equiv \rho_{\sin}\left(s_1, \mathbf{x}^{(1)}_{N_1}; \cdots ; s_M, \mathbf{x}^{(M)}_{N_M}\right), \tag{3.120}$$

where

$$\mathbf{K}_{\sin}(s,x;t,y) = \begin{cases} \displaystyle\int_0^1 du\, e^{-\pi^2 u^2 (s-t)} \cos\{\pi u(x-y)\}, & \text{if } t>s, \\ \mathrm{K}_{\sin}(x,y), & \text{if } t=s, \\ \displaystyle -\int_1^\infty du\, e^{-\pi^2 u^2 (s-t)} \cos\{\pi u(x-y)\}, & \text{if } t<s, \end{cases} \tag{3.121}$$

with

$$\mathrm{K}_{\sin}(x,y) = \frac{1}{2\pi}\int_{|k|\le\pi} dk\, e^{-k^2 t/2+\sqrt{-1}k(x-y)} = \frac{\sin\{\pi(x-y)\}}{\pi(x-y)}, \quad x,y\in\mathbb{R}. \tag{3.122}$$

Proof For any $u \in \mathbb{R}$, the formulas

$$\lim_{\ell\to\infty} (-1)^\ell \ell^{1/4} \varphi_{2\ell}\left(\frac{u}{2\sqrt{\ell}}\right) = \frac{1}{\sqrt{\pi}}\cos u,$$
$$\lim_{\ell\to\infty} (-1)^\ell \ell^{1/4} \varphi_{2\ell+1}\left(\frac{u}{2\sqrt{\ell}}\right) = \frac{1}{\sqrt{\pi}}\sin u \tag{3.123}$$

are known (see Eq. (8.22.8) in Chap. 8 of [129]). We note that

$$\left(\frac{t_n}{t_m}\right)^\alpha = \left(\frac{N/\pi^2+2s_n}{N/\pi^2+2s_m}\right)^\alpha$$
$$= \left\{\left(1+\frac{2\pi^2 s_n}{N}\right)^N \left(1+\frac{2\pi^2 s_m}{N}\right)^{-N}\right\}^{\alpha/N}$$
$$\simeq e^{-2\pi^2\alpha(s_m-s_n)/N}$$

for $N \gg 1$ with a fixed number α. Then (3.112) with $s = t_m = N/\pi^2 + 2s_m \le t = t_n = N/\pi^2 + 2s_n$ is evaluated at $N \to \infty$ as

$$
\begin{aligned}
\mathbf{K}^{(N)}_{\text{Hermite}}(t_m, x; t_n, y) &\simeq \frac{1}{N} \sum_{\ell=0}^{N/2-1} e^{-2\pi^2 \ell (s_m - s_n)/N} \\
&\times \sqrt{\frac{N}{2\ell}} \left\{ \cos\left(\pi \sqrt{\frac{2\ell}{N}} x\right) \cos\left(\pi \sqrt{\frac{2\ell}{N}} y\right) + \sin\left(\pi \sqrt{\frac{2\ell}{N}} x\right) \sin\left(\pi \sqrt{\frac{2\ell}{N}} y\right) \right\} \\
&\simeq \frac{1}{2} \int_0^1 \frac{d\lambda}{\sqrt{\lambda}} e^{-\pi^2 \lambda (s_m - s_n)} \left\{ \cos(\pi\sqrt{\lambda} x) \cos(\pi \sqrt{\lambda} y) + \sin(\pi \sqrt{\lambda} x) \sin(\pi \sqrt{\lambda} y) \right\} \\
&= \int_0^1 du \, e^{-\pi^2 u^2 (s_m - s_n)} \cos\{\pi u (x - y)\}.
\end{aligned}
$$

In particular, when $t_m = t_n$, i.e., $s_n - s_m = 0$, the integration is readily performed to have $\int_0^1 du \, \cos\{\pi u(x-y)\} = \sin\{\pi(x-y)\}/\pi(x-y)$. A similar evaluation at $N \to \infty$ can be done also for (3.112) with $s = t_m > t = t_n$. □

The correlation kernel (3.121) is called the *extended sine kernel*. Since it is a function of $s - t$ and $x - y$, the determinantal process obtained by the bulk scaling limit is a temporally and spatially homogeneous process with an infinite number of particles, which we write as $(\Xi, \mathbf{P}_{\text{sin}})$. Let P_{sin} be a stationary probability measure on $\mathbb{R}$, which is a determinantal point process [123, 124, 126] such that the spatial correlation function is given by

$$
\rho_{\text{sin}}(\mathbf{x}_N) = \det_{1 \le i,j \le N} \left[\text{K}_{\text{sin}}(x_i, x_j) \right] \tag{3.124}
$$

for any $N \in \mathbb{N}$, $\mathbf{x}_N = (x_1, \dots, x_N) \in \mathbb{R}^N$, where K_{sin} is given by (3.122). The determinantal process $(\Xi, \mathbf{P}_{\text{sin}})$ is reversible with respect to P_{sin}.

3.9.3 Soft-Edge Scaling Limit and Spatially Inhomogeneous Infinite System

Next we consider the scaling limit

$$
t \simeq N^{1/3} \quad \text{and} \quad x \simeq 2N^{2/3}. \tag{3.125}
$$

Since (3.125) gives $x^2/2t \simeq 2N$, this scaling limit allows us to zoom into the right edge of the semicircle-shaped profile (3.117), and we obtain a spatially inhomogeneous infinite particle system. Following random matrix theory [101], we call (3.125) the *soft-edge scaling limit*.

In order to describe the limit, we introduce the *Airy function*

$$\mathrm{Ai}(x) = \frac{1}{\pi} \int_0^{\infty} dk \, \cos\left(\frac{k^3}{3} + kx \right). \tag{3.126}$$

It is the solution of the equation

$$\frac{d^2}{dx^2} \mathrm{Ai}(x) = x \mathrm{Ai}(x), \tag{3.127}$$

which obeys the asymptotics given by

$$\begin{aligned} \mathrm{Ai}(x) &\simeq \frac{1}{2\sqrt{\pi} x^{1/4}} \exp\left(-\frac{2}{3} x^{3/2} \right), \\ \mathrm{Ai}(-x) &\simeq \frac{1}{\sqrt{\pi} x^{1/4}} \cos\left(\frac{2}{3} x^{3/2} - \frac{\pi}{4} \right) \quad \text{as } x \to \infty. \end{aligned} \tag{3.128}$$

In the proof of the following theorem, we will use the formula

$$\lim_{\ell \to \infty} 2^{-1/4} \ell^{1/12} \varphi_\ell \left(\sqrt{2\ell} + \frac{x}{\sqrt{2}} \ell^{-1/6} \right) = \mathrm{Ai}(x) \quad \text{for } x \in \mathbb{R}, \tag{3.129}$$

which is obtained from Theorem 8.22.9 (c) in Chap. 8 of [129]. Let

$$a_N(s) = 2N^{2/3} + N^{1/3} s - \frac{s^2}{4}, \tag{3.130}$$

and $a_N(s) + \mathbf{x}_{N'} = (a_N(s) + x_1, a_N(s) + x_2, \cdots, a_N(s) + x_{N'})$.

Proposition 3.7 *For any $M \in \mathbb{N}$, any sequence $\{N_m\}_{m=1}^M$ of positive integers, and any strictly increasing sequence $\{s_m\}_{m=1}^M$ of positive numbers*

$$\begin{aligned} &\lim_{N \to \infty} \rho^{(N)}_{\mathrm{Hermite}} \left(N^{1/3} + s_1, a_N(s_1) + \mathbf{x}^{(1)}_{N_1}; \cdots ; N^{1/3} + s_M, a_N(s_M) + \mathbf{x}^{(M)}_{N_M} \right) \\ &= \det_{\substack{1 \le i \le N_m, 1 \le j \le N_n, \\ 1 \le m, n \le M}} \left[\mathbf{K}_{\mathrm{Airy}}(s_m, x^{(m)}_i; s_n, x^{(n)}_j) \right] \\ &\equiv \rho_{\mathrm{Airy}} \left(s_1, \mathbf{x}^{(1)}_{N_1}; \cdots ; s_M, \mathbf{x}^{(M)}_{N_M} \right), \end{aligned} \tag{3.131}$$

where

$$\mathbf{K}_{\mathrm{Airy}}(s, x; t, y) = \begin{cases} \displaystyle\int_0^{\infty} du \, e^{u(s-t)/2} \mathrm{Ai}(x+u) \mathrm{Ai}(y+u), & \text{if } t \ge s, \\ \displaystyle -\int_{-\infty}^{0} du \, e^{u(s-t)/2} \mathrm{Ai}(x+u) \mathrm{Ai}(y+u), & \text{if } t < s. \end{cases} \tag{3.132}$$

Proof Replacing the summation index in (3.112) by $N-p-1$ for the case where $m \le n$, we have

$$\mathbb{K}_N(t_m, x; t_n, y) = \left(\frac{t_n}{t_m}\right)^{(N-1)/2} \frac{1}{\sqrt{2t_m}} \sum_{p=0}^{N-1} \left(\frac{t_n}{t_m}\right)^{-p/2} \varphi_{N-p-1}\left(\frac{x}{\sqrt{2t_m}}\right) \varphi_{N-p-1}\left(\frac{y}{\sqrt{2t_n}}\right).$$

When we set $t_m = N^{1/3} + s_m$, we see that

$$\frac{a_N(s_m)+x}{\sqrt{2t_m}} = \sqrt{2N} + \frac{x}{\sqrt{2}} N^{-1/6} + \mathcal{O}(N^{-1/2}), \tag{3.133}$$

and we can use the formula (3.129):

$$\begin{aligned}
\varphi_{N-p-1}\left(\frac{x}{\sqrt{2t_m}}\right) &\simeq \varphi_{N-p-1}\left(\sqrt{2N} + \frac{x}{\sqrt{2}} N^{-1/6}\right) \\
&\simeq \varphi_{N-p-1}\left(\sqrt{2(N-p-1)} + \frac{1}{\sqrt{2}}(N-p-1)^{-1/6}\left\{x + \frac{p}{N^{1/3}}\right\}\right) \\
&\simeq 2^{1/4} N^{-1/12} \mathrm{Ai}\left(x + \frac{p}{N^{1/3}}\right).
\end{aligned}$$

For

$$\left(\frac{t_n}{t_m}\right)^{-p/2} = \left[\left(\frac{1+s_n/N^{1/3}}{1+s_m/N^{1/3}}\right)^{N^{1/3}/2}\right]^{-p/N^{1/3}} \simeq e^{p(s_m-s_n)/2N^{1/3}} \quad \text{as } N \to \infty,$$

we have, for $n \ge m$,

$$\begin{aligned}
&\mathbf{K}^{(N)}_{\text{Hermite}}(N^{1/3}+s_m, a_N(s_m)+x; N^{1/3}+s_n, a_N(s_n)+y) \\
&\sim \frac{1}{N^{1/3}} \sum_{p=0}^{N-1} e^{p(s_m-s_n)/2N^{1/3}} \mathrm{Ai}\left(x+\frac{p}{N^{1/3}}\right) \mathrm{Ai}\left(y+\frac{p}{N^{1/3}}\right) \\
&\simeq \int_0^\infty du\, e^{u(s_m-s_n)/2} \mathrm{Ai}(x+u)\mathrm{Ai}(y+u) \quad \text{as } N \to \infty.
\end{aligned}$$

Note that the factor $(t_n/t_m)^{(N-1)/2}$ was omitted in the second line in the above equations, since it is irrelevant in calculating determinants by Lemma 3.8. A similar evaluation at $N \to \infty$ of (3.112) can be done also for $m > n$. □

The infinite system obtained by the soft-edge scaling limit (3.125) is temporally homogeneous, but spatially inhomogeneous as shown by the correlation kernel $\mathbf{K}_{\text{Airy}}$, (3.132). We call $\mathbf{K}_{\text{Airy}}$ the *extended Airy kernel* [77, 106] and write this stationary determinantal process as $(\Xi, \mathbf{P}_{\text{Airy}})$. Prähofer and Spohn [114] and Johansson [65] studied the rightmost path in the present system and called it the *Airy process* $(A(t))_{t\geq 0}$. For a given $t > 0$, $A(t)$ follows the celebrated *Tracy–Widom distribution*, which is governed by the Painlevé II equation as shown in Sect. 3.11 [136, 137]. (See [62] for the original work on the relationship between the Painlevé transcendent and interacting particle systems.) Tracy and Widom also derived a system of PDEs, which govern $(A(t))_{t\geq 0}$ [139]. See also [2, 42, 140].

Let P_{Airy} be the stationary probability measure on $\mathbb{R}$, which is a determinantal point process [123, 124, 126] such that the spatial correlation function is given by

$$\rho_{\text{Airy}}(\mathbf{x}_N) = \det_{1\leq i,j\leq N}\Big[\mathrm{K}_{\text{Airy}}(x_i, x_j)\Big] \tag{3.134}$$

for any $N \in \mathbb{N}, \mathbf{x}_N = (x_1, \dots, x_N) \in \mathbb{R}^N$, where

$$\begin{aligned} \mathrm{K}_{\text{Airy}}(x, y) &= \mathbf{K}_{\text{Airy}}(t, x; t, y) \\ &= \int_0^\infty du\, \mathrm{Ai}(x+u)\mathrm{Ai}(y+u). \end{aligned} \tag{3.135}$$

The *Airy kernel* K_{Airy} is also written as (Exercise 3.18)

$$\mathrm{K}_{\text{Airy}}(x, y) = \frac{\mathrm{Ai}(x)\mathrm{Ai}'(y) - \mathrm{Ai}'(x)\mathrm{Ai}(y)}{x-y}, \quad x \neq y, \tag{3.136}$$

$$\mathrm{K}_{\text{Airy}}(x, x) = \mathrm{Ai}'(x)^2 - x\mathrm{Ai}(x)^2. \tag{3.137}$$

3.10 Entire Functions and Infinite Particle Systems

A function which is represented by a power series of the form

$$f(z) = \sum_{n=0}^{\infty} c_n z^n, \quad z \in \mathbb{C}$$

with $\lim_{n\to\infty} |c_n|^{1/n} = 0$, is analytic in the whole complex plane and is called an *entire function*. The class of entire functions includes all polynomials. As polynomials are classified by their degree indicating their growth as $|z| \to \infty$, entire functions are classified by the *order of growth* ρ_f defined by

$$\rho_f = \limsup_{r\to\infty} \frac{\log\log M_f(r)}{\log r} \quad \text{for } M_f(r) = \max_{|z|=r} |f(z)|. \tag{3.138}$$

Note that (3.138) means $\max_{|z|=r} |f(z)| \sim \exp(\sigma_f z^{\rho_f})$ as $z \to \infty$, where the constant σ_f is called the *type*: $\sigma_f \equiv \limsup_{r\to\infty} \log M_f(r)/r^{\rho_f}$.

For $\xi = \sum_{i=1}^N \delta_{x_i} \in \mathfrak{M}_0$ with $p \in \mathbb{N}_0$, we consider the product

$$\Pi_p(\xi, z) = \prod_{1\le i\le N, x_i\neq 0} G\left(\frac{z}{x_i}, p\right), \quad z \in \mathbb{C},$$

where

$$G(u,p) = \begin{cases} 1-u, & \text{if} \quad p=0, \\ (1-u)\exp\left[u + \dfrac{u^2}{2} + \cdots + \dfrac{u^p}{p}\right], & \text{if} \quad p \in \mathbb{N}. \end{cases} \tag{3.139}$$

The function $G(u, p)$ is called the *Weierstrass primary factor* with *genus* p. With $\alpha > 0$, we consider

$$\zeta_\xi(\alpha) = \int_{\{0\}^c} \frac{1}{|x|^\alpha} \xi(dx) = \sum_{i, x_i \neq 0} \frac{1}{|x_i|^\alpha}. \tag{3.140}$$

For a configuration with an infinite number of particles, $\xi = \sum_i \in \mathfrak{M}_0$ with $\xi(\mathbb{R}) = \infty$, we put $\xi \cap [-L, L] \equiv \sum_{i, x_i \in [-L,L]} \delta_{x_i}$ for $L > 0$. We define

$$\zeta_\xi(\alpha) = \lim_{L\to\infty} \zeta_{\xi\cap[-L,L]}(\alpha), \tag{3.141}$$

if the limit finitely exists. If $\zeta_\xi(p+1) < \infty$ for some $p \in \mathbb{N}_0$, the limit

$$\Pi_p(\xi, z) = \lim_{L\to\infty} \Pi_p(\xi \cap [-L, L], z) = \prod_{i, x_i\neq 0} G\left(\frac{z}{x_i}, p\right), \quad z \in \mathbb{C}, \tag{3.142}$$

finitely exists. This infinite product is called the *Weierstrass canonical product of genus* p [92].

The *Hadamard theorem* [92] claims that any entire function f of finite order $\rho_f < \infty$ can be represented by

$$f(z) = z^m e^{P_q(z)} \Pi_p(\xi_f, z), \tag{3.143}$$

where the genus p satisfies the inequality

$$p \le \rho_f. \tag{3.144}$$

Here $P_q(z)$ is a polynomial in z of degree $q \le \rho_f$, m is the multiplicity of the root at the origin, and ξ_f is the distribution of zeros of the entire function f except for the origin:

$$\xi_f = \sum_{x \in f^{-1}(0), x \neq 0} \delta_x. \tag{3.145}$$

The relation (3.144) between the genus p and the order of growth ρ_f implies that, if an entire function f grows very rapidly, then its zeros are arranged very densely. One of the main subjects of the theory of entire functions is to clarify the relationship between the growth of an entire function and the distribution of its zeros. We use this theory to construct and analyze determinantal processes with an infinite number of particles [83–85].

3.10.1 Nonequilibrium Sine Process

The function $f(z) = \sin(\pi\sqrt{z})/\pi\sqrt{z}$ is entire and of order $\rho_f = 1/2$ with zeros $f(0)^{-1} = \{n^2 : n \in \mathbb{N}\}$. According to the Hadamard theorem, we have

$$\frac{\sin(\pi\sqrt{z})}{\pi\sqrt{z}} = \prod_{n=1}^{\infty}\left(1 - \frac{z}{n^2}\right), \quad z \in \mathbb{C},$$

since $f(0) = 1$. By replacing z by z^2, we obtain

$$\sin(\pi z) = \pi z \prod_{\substack{n \in \mathbb{Z}, \\ n \neq 0}} \left(1 - \frac{z}{n}\right), \quad z \in \mathbb{C}. \tag{3.146}$$

Define

$$\begin{aligned} \Phi^{\ell}_{\xi^{\mathbb{Z}}}(z) &\equiv \prod_{\substack{n \in \mathbb{Z}, \\ n \neq \ell}} \frac{z-n}{\ell-n} = \frac{\sin\{\pi(z-\ell)\}}{\pi(z-\ell)} \\ &= \frac{1}{\pi(z-\ell)} \frac{\sin(\pi z)}{\sin'(\pi\ell)}, \quad z \in \mathbb{C}, \quad \ell \in \mathbb{Z}, \end{aligned} \tag{3.147}$$

where the second equality is given by (3.146) and the third one is verified by noting that $\sin\{\pi(z-\ell)\} = \sin(\pi z)\cos(\pi\ell)$ and $\sin'(\pi\ell) = \cos(\pi\ell) = (-1)^{\ell}$ for $\ell \in \mathbb{Z}$. It can be regarded as an extension of the polynomial given by (3.60) to an entire function $\Phi^{\ell}_{\xi^{\mathbb{Z}}}(z)$, $z \in \mathbb{C}$, where $\ell \in \mathbb{Z}$ and $\xi^{\mathbb{Z}}$ is the configuration in which every point of $\mathbb{Z}$ is occupied by one particle:

$$\xi^{\mathbb{Z}}(\cdot) = \sum_{n \in \mathbb{Z}} \delta_n(\cdot). \tag{3.148}$$

(The last expression of $\Phi^{\ell}_{\xi\mathbb{Z}}(z)$ in (3.147) will be compared with the definition of $\widehat{\Psi}^{a_\ell}_{\xi\mathscr{A}}(z)$ given by the first equality in (3.176) [83].) The function (3.147) has the following integral representation,

$$\Phi^{\ell}_{\xi\mathbb{Z}}(z) = \frac{1}{2\pi}\int_{|k|\leq\pi} e^{\sqrt{-1}k(z-\ell)}dk, \quad z\in\mathbb{C}, \quad \ell\in\mathbb{Z}. \tag{3.149}$$

The martingale function (3.62) with (1.25) corresponding to (3.149) will be

$$\begin{aligned}
\mathscr{M}^{\ell}_{\xi\mathbb{Z}}(t,y) &= \mathscr{I}[\Phi^{\ell}_{\xi\mathbb{Z}}(W)|(t,y)] \\
&= \int_{\mathbb{R}} dw\, \Phi^{\ell}_{\xi\mathbb{Z}}(\sqrt{-1}w)\frac{e^{-(\sqrt{-1}y+w)^2/2t}}{\sqrt{2\pi t}} \\
&= \frac{1}{2\pi}\int_{|k|\leq\pi} dk \int_{\mathbb{R}} dw\, e^{\sqrt{-1}k(\sqrt{-1}w-\ell)}\frac{e^{-(\sqrt{-1}y+w)^2/2t}}{\sqrt{2\pi t}} \\
&= \frac{1}{2\pi}\int_{|k|\leq\pi} dk\, e^{tk^2/2+\sqrt{-1}k(y-\ell)}.
\end{aligned} \tag{3.150}$$

Since the integrand of (3.150) is equal to $G_{\sqrt{-1}k}(t, y-\ell)$, where $G_\alpha(t,x)$ is given by (1.15), $\mathscr{M}^{\ell}_{\xi\mathbb{Z}}(t, B(t)),\ t\geq 0,\ \ell\in\mathbb{Z}$ are indeed martingales. Then following (3.89), we put

$$\begin{aligned}
\mathscr{G}_{\xi\mathbb{Z}}(s,x;t,y) &= \sum_{\ell\in\mathbb{Z}} p(s,x|\ell)\mathscr{M}^{\ell}_{\xi Z}(t,y) \\
&= \frac{1}{2\pi}\int_{|k|\leq\pi} dk\, \frac{1}{\sqrt{2\pi s}}\sum_{\ell\in\mathbb{Z}} e^{-(x-\ell)^2/2s+tk^2/2+\sqrt{-1}k(y-\ell)}.
\end{aligned} \tag{3.151}$$

Here we introduce the *Jacobi theta function* defined by

$$\vartheta_3(z,\tau) = \sum_{\ell\in\mathbb{Z}} e^{2\pi\sqrt{-1}z\ell+\pi\sqrt{-1}\tau\ell^2}, \quad z\in\mathbb{C}, \quad \Im\tau>0. \tag{3.152}$$

Then (3.151) is written as

$$\begin{aligned}
&\frac{1}{2\pi}\int_{|k|\leq\pi} dk\, e^{k^2(t-s)/2+\sqrt{-1}k(y-x)}\vartheta_3\left(\frac{1}{2\pi\sqrt{-1}s}(x-\sqrt{-1}ks), -\frac{1}{2\pi\sqrt{-1}s}\right) \\
&\quad\times e^{-\pi\sqrt{-1}(x-\sqrt{-1}ks)^2/2\pi\sqrt{-1}s}\sqrt{\frac{\sqrt{-1}}{2\pi\sqrt{-1}s}}.
\end{aligned}$$

We use the functional equation satisfied by $\vartheta_3(z, \tau)$,

$$\vartheta_3(z, \tau) = \vartheta_3\left(\frac{z}{\tau}, -\frac{1}{\tau}\right) e^{-\pi\sqrt{-1}z^2/\tau} \sqrt{\frac{\sqrt{-1}}{\tau}}, \quad z \in \mathbb{C}, \quad \Im\tau > 0,$$

which is called *Jacobi's imaginary transformation* (see, for example, Sect. 10.12 in [8]). Then we have

$$\mathscr{G}_{\xi^{\mathbb{Z}}}(s, x; t, y) = \frac{1}{2\pi} \int_{|k| \leq \pi} dk\, e^{k^2(t-s)/2+\sqrt{-1}k(y-x)} \vartheta_3(x - \sqrt{-1}ks, 2\pi\sqrt{-1}s).$$

Following (3.88), we set

$$\mathbb{K}_{\xi^{\mathbb{Z}}}(s, x; t, y) = \mathscr{G}_{\xi^{\mathbb{Z}}}(s, x; t, y) - \mathbf{1}_{(s>t)} p(s-t, x|y), \quad (s, x), (t, y) \in [0, \infty) \times \mathbb{R}. \tag{3.153}$$

By the procedure above, it is expected that this gives the correlation kernel of the determinantal process starting from the configuration (3.148) with an infinite number of particles. See Fig. 3.1. This is a fact, and the following statement was proved in [84].

Theorem 3.6 *Consider the system of SDEs for the Dyson model with an infinite number of particles,*

$$dX_i(t) = dB_i(t) + \sum_{\substack{j \in \mathbb{Z}, \\ j \neq i}} \frac{dt}{X_i(t) - X_j(t)}, \quad t \geq 0, \quad i \in \mathbb{Z}. \tag{3.154}$$

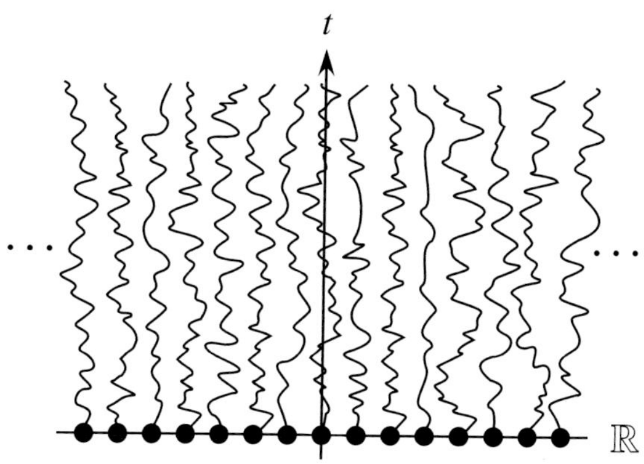

Fig. 3.1 Consider the Dyson model starting from the configuration in which every point of the integers $\mathbb{Z}$ is occupied by one particle. This nonequilibrium determinantal process shows a relaxation phenomenon to the stationary state $\mathrm{P}_{\sin}$

For the initial configuration $\xi^{\mathbb{Z}}$ given by (3.148), the solution of this infinite system of SDEs is given by a determinantal process $(\Xi, \mathbb{P}^{\xi^{\mathbb{Z}}})$ whose correlation kernel is given by (3.153).

The correlation kernel (3.153) is divided into two parts,

$$\mathbb{K}_{\xi^{\mathbb{Z}}}(s, x; t, y) = \mathbf{K}_{\sin}(s, x; t, y) + R_{\xi^{\mathbb{Z}}}(s, x; t, y), \tag{3.155}$$

where $\mathbf{K}_{\sin}$ is the extended sine kernel given by (3.121) and

$$\begin{aligned}
&R_{\xi^{\mathbb{Z}}}(s, x; t, y)\\
&= \frac{1}{2\pi}\int_{|k|\le\pi} dk\, e^{k^2(t-s)/2+\sqrt{-1}k(y-x)}\left\{\vartheta_3(x-\sqrt{-1}ks, 2\pi\sqrt{-1}s) - 1\right\}\\
&= \sum_{n\in\mathbb{N}} e^{-2\pi^2 sn^2}\frac{1}{\pi}\int_{|k|\le\pi} dk\, e^{k^2(t-s)/2+2\pi ksn}\cos[2\pi xn + k(x-y)],
\end{aligned}$$

$(s, x), (t, y) \in [0, \infty) \times \mathbb{R}$. For any $T > 0$, we can see that

$$\begin{aligned}
&\left|R_{\xi^{\mathbb{Z}}}(s+T, x; t+T, y)\right|\\
&\le \left(e^{\pi^2(t-s)/2} \vee 1\right)\sum_{n\in\mathbb{N}} e^{-2\pi^2(s+T)n^2}\frac{1}{\pi}\int_{|k|\le\pi} dk\, e^{2\pi k(s+T)n}\\
&\le \frac{e^{\pi^2(t-s)/2}\vee 1}{2\pi^2(s+T)} \le \frac{C}{T}, \quad (s, x), (t, y) \in [0, \infty) \times \mathbb{R},
\end{aligned}$$

where $a \vee b = \max\{a, b\}$. Here $C > 0$ depends on s and t, but not on T. Then we can conclude that for any fixed $s, t > 0$,

$$\lim_{T\to\infty} \mathbb{K}_{\xi^{\mathbb{Z}}}(s+T, x; t+T, y) = \mathbf{K}_{\sin}(s, x; t, y) \tag{3.156}$$

uniformly on any compact subset of $\mathbb{R}^2$. The convergence of the correlation kernel implies the following.

Proposition 3.8 *The Dyson model (3.154) with an infinite number of particles starting from $\xi^{\mathbb{Z}}$, $(\Xi, \mathbb{P}^{\xi^{\mathbb{Z}}})$, shows a relaxation phenomenon to the determinantal process $(\Xi, \mathbf{P}_{\sin})$, which is specified by the extended sine kernel $\mathbf{K}_{\sin}$ obtained by the bulk scaling limit from the extended Hermite kernel.*

A *relaxation phenomenon* is a typical phenomenon observed in nonequilibrium dynamics and we call the determinantal infinite-particle system $(\Xi, \mathbb{P}^{\xi^{\mathbb{Z}}})$ specified by $\mathbb{K}^{\xi^{\mathbb{Z}}}$ the *nonequilibrium sine process*. The determinantal process $(\Xi, \mathbf{P}_{\sin})$ with the correlation kernel $\mathbf{K}_{\sin}(s, x; t, y)$ is the *equilibrium dynamics*, which is reversible with respect to $P_{\sin}$. The relaxation exhibits slow dynamics. As a matter of fact, we can find that the particle density function $\rho_{\xi^{\mathbb{Z}}}(t, x)$ shows a *power-law behavior*

$$\rho_{\xi^{\mathbb{Z}}}(t, x) - 1 = \frac{1}{\pi^2 t} \sum_{n \in \mathbb{N}} \frac{\cos(2\pi n x)}{n} \sinh(2\pi^2 n t) e^{-2\pi^2 n^2 t}$$
$$\simeq \frac{\cos(2\pi x)}{2\pi^2} \frac{1}{t}, \quad x \in \mathbb{R}, \quad t \to \infty, \tag{3.157}$$

in approaching the uniform equilibrium state $\rho_{\sin}(x) \equiv 1$.

3.10.2 *Nonequilibrium Airy Process*

The Airy function defined by (3.126) as a real function allows us to define it as an entire function by

$$\mathrm{Ai}(z) = \frac{1}{2\pi\sqrt{-1}} \int_C e^{k^3/3 - zk} dk, \quad z \in \mathbb{C}, \tag{3.158}$$

where the contour C starts at $e^{-2\pi\sqrt{-1}/3}\infty$ and finishes at $e^{2\pi\sqrt{-1}/3}\infty$, following the corresponding rays asymptotically, staying in the sector $-2\pi/3 < \arg z < 2\pi/3$ [142]. As suggested by (3.128), the order of growth is

$$\rho_{\mathrm{Ai}} = \frac{3}{2} \tag{3.159}$$

with type $\sigma_{\mathrm{Ai}} = 2/3$;

$$\max_{|z|=r} |\mathrm{Ai}(z)| \sim \exp\left(\frac{2}{3} r^{3/2}\right) \quad \text{as } r \to \infty. \tag{3.160}$$

The zeros of the Airy function are located only on the negative part of the real axis $\mathbb{R}$,

$$\mathscr{A} \equiv \mathrm{Ai}^{-1}(0) = \left\{ a_n, n \in \mathbb{N} : \mathrm{Ai}(a_n) = 0,\ 0 > a_1 > a_2 > \cdots \right\}, \tag{3.161}$$

with the values [1] $a_1 = -2.338\ldots$, $a_2 = -4.087\ldots$, $a_3 = -5.520\ldots$, $a_4 = -6.786\ldots$, $a_5 = -7.944\ldots$, and they admit the asymptotics [1, 142]

$$a_n \simeq -\left(\frac{3\pi}{2}\right)^{2/3} n^{2/3} \quad \text{as } n \to \infty. \tag{3.162}$$

Then the average density of zeros of the Airy function around x, denoted by $\rho_{\mathscr{A}}(x)$, behaves as

$$\rho_{\mathscr{A}}(x) \simeq \frac{1}{\pi}(-x)^{1/2} \to \infty \quad \text{as } x \to -\infty. \tag{3.163}$$

Let

$$\xi^{\mathscr{A}} = \sum_{a \in \mathscr{A}} \delta_a = \sum_{n=1}^{\infty} \delta_{a_n} \in \mathfrak{M}_0. \tag{3.164}$$

By (3.163), the 'Airy zeta function' [142] gives $\zeta_{\xi^{\mathscr{A}}}(1) = \sum_{a \in \mathscr{A}} 1/|a| = \infty$, but

$$\zeta_{\xi^{\mathscr{A}}}(2) = \sum_{a \in \mathscr{A}} \frac{1}{|a|^2} = d_1^2 < \infty \tag{3.165}$$

with

$$d_1 = \frac{\mathrm{Ai}'(0)}{\mathrm{Ai}(0)} = -\frac{3^{1/3}\Gamma(2/3)}{\Gamma(1/3)} = -\frac{3^{5/6}(\Gamma(2/3))^2}{2\pi} = -0.7290\ldots, \tag{3.166}$$

where $\mathrm{Ai}'(z) = d\mathrm{Ai}(z)/dz$. According to the Hadamard theorem, $\mathrm{Ai}(z)$ is expressed using the Weierstrass canonical product of genus $p = 1$ as

$$\begin{aligned} \mathrm{Ai}(z) &= e^{d_0 + d_1 z} \Pi_1(\xi^{\mathscr{A}}, z) \\ &= e^{d_0 + d_1 z} \prod_{n=1}^{\infty} \left\{ \left(1 - \frac{z}{a_n}\right) e^{z/a_n} \right\}, \quad z \in \mathbb{C}, \end{aligned} \tag{3.167}$$

where

$$d_0 = \log \mathrm{Ai}(0) = -\log(3^{2/3}\Gamma(2/3)) = 1.035\ldots. \tag{3.168}$$

First we show that the Airy function is related to the drifted Brownian motion of the form

$$B(t) + \frac{t^2}{4}, \quad t \geq 0. \tag{3.169}$$

The transition probability density is given by

$$\begin{aligned} p_{t^2/4}(t, y|s, x) &\equiv p(t-s, y - t^2/4 | x - s^2/4) \\ &= \frac{1}{\sqrt{2\pi(t-s)}} \exp\left[-\frac{(y-x)^2}{2(t-s)} + \frac{(t+s)(y-x)}{4} - \frac{(t-s)(t+s)^2}{32} \right]. \end{aligned} \tag{3.170}$$

Let

$$p_{\mathrm{Ai}}(t, y|x) \equiv \int_{\mathbb{R}} du\, e^{ut/2} \mathrm{Ai}(x+u)\mathrm{Ai}(y+u), \quad x, y \in \mathbb{R}, \quad t \geq 0, \tag{3.171}$$

and

$$g(t, x) \equiv \int_{\mathbb{R}} dz\, p_{\mathrm{Ai}}(t, z|x) = \exp\left(-\frac{tx}{2} + \frac{t^3}{24}\right). \tag{3.172}$$

Then the following relation is established,

$$p_{\mathrm{Ai}}(t-s, y|x) = \frac{g(t, y)}{g(s, x)} p_{t^2/4}(t, y|s, x), \quad x, y \in \mathbb{R}, \quad 0 \leq s \leq t. \tag{3.173}$$

When $s = 0$, $g(0, x) = 1$ and (3.173) becomes

$$p_{\mathrm{Ai}}(t, y|x) = g(t, y) p_{t^2/4}(t, y|0, x), \quad x, y \in \mathbb{R}, \quad t \geq 0. \tag{3.174}$$

Corresponding to the drift (3.169), the kernel of the integral transformation $\widehat{G}_w(t, x)$ given by (1.22) is replaced by $\widehat{G}_w(t, x - t^2/4)$ and we consider the integral transformation

$$\mathscr{I}[f(W)|(t, x - t^2/4)] = \int_{-\infty}^{\infty} dw\, f(\sqrt{-1}w)\widehat{G}_w(t, x - t^2/4). \tag{3.175}$$

Now we consider the nonequilibrium determinantal process with an infinite number of particles starting from the Airy zeros (3.164). We write it as $(\Xi, \mathbb{P}^{\xi^{\mathscr{A}}})$. The following one-parameter family of entire functions $\{\widehat{\Psi}^{a_\ell}_{\xi^{\mathscr{A}}}(z)\}_{\ell \in \mathbb{N}}$ is considered:

$$\begin{aligned}
\widehat{\Psi}^{a_\ell}_{\xi^{\mathscr{A}}}(z) &\equiv \frac{1}{z - a_\ell} \frac{\mathrm{Ai}(z)}{\mathrm{Ai}'(a_\ell)} = e^{d_1(z-a_\ell)} \prod_{m=1}^{\infty} e^{(z-a_\ell)/a_m} \prod_{\substack{n \in \mathbb{N}, \\ n \neq \ell}} \frac{z - a_n}{a_\ell - a_n} \\
&= \exp\left[\left(d_1 + \sum_{n=1}^{\infty} \frac{1}{a_n}\right)(z - a_\ell)\right] \widehat{\Phi}^{a_\ell}_{\xi^{\mathscr{A}}}(z), \quad \ell \in \mathbb{N}, \ z \in \mathbb{C},
\end{aligned} \tag{3.176}$$

where

$$\widehat{\Phi}^x_\xi(z) = \prod_{\substack{n \in \mathbb{N}, \\ x_n \neq x}} \frac{z - x_n}{x - x_n} \quad \text{for } \xi = \sum_{n \in \mathbb{N}} \delta_{x_n}. \tag{3.177}$$

The function (3.176) seems to be similar to the integrand $e^{b(W-x)}\Phi^x_\xi(W)$ found in (3.93), but we should note that

$$d_1 + \sum_{n=1}^{N} \frac{1}{a_n} \simeq -\left(\frac{12}{\pi^2}\right)^{1/3} N^{1/3} \to -\infty \quad \text{as } N \to \infty, \tag{3.178}$$

and $\widehat{\Phi}^x_{\xi\mathscr{A}}(z) = \infty$. Nevertheless, $\{\widehat{\Psi}^{a_\ell}_{\xi\mathscr{A}}(z)\}_{\ell\in\mathbb{N}}$ defined by (3.176) are entire functions. The one-parameter family of martingale functions $\{\mathscr{M}^{a_\ell}_{\xi\mathscr{A}}(t,y)\}_{\ell\in\mathbb{N}}$ should be obtained by

$$\begin{aligned}\mathscr{M}^{a_\ell}_{\xi\mathscr{A}}(t,y) &= \mathscr{I}\left[\widehat{\Psi}^{a_\ell}_{\xi\mathscr{A}}(W)\Big|(t, y-t^2/4)\right] \\ &= \mathscr{I}\left[e^{[d_1+\sum_{n=1}^{\infty}(1/a_n)](W-a_\ell)}\widehat{\Phi}^{a_\ell}_{\xi\mathscr{A}}(W)\Big|(t,y-t^2/4)\right], \end{aligned} \tag{3.179}$$

$\ell \in \mathbb{N},\ y \in \mathbb{R},\ t \geq 0.$

The following integral representations of the entire functions are available (Exercise 3.19),

$$\widehat{\Psi}^{a_\ell}_{\xi\mathscr{A}}(z) = \frac{1}{\mathrm{Ai}'(a_\ell)^2} \int_0^\infty du\, \mathrm{Ai}(z+u)\mathrm{Ai}(a_\ell+u) \tag{3.180}$$

for $a_\ell \in \mathscr{A}, z \neq a_\ell$. Then

$$\mathscr{M}^{a_\ell}_{\xi\mathscr{A}}(t,y) = \frac{1}{\mathrm{Ai}'(a_\ell)^2} \int_0^\infty du\, \mathrm{Ai}(a_\ell+u)\mathscr{I}[\mathrm{Ai}(W+u)|(t,y-t^2/4)].$$

Here we can prove that (Exercise 3.20)

$$\mathscr{I}[\mathrm{Ai}(W+u)|(t,y-t^2/4)] = g(t,y)e^{-ut/2}\mathrm{Ai}(y+u), \tag{3.181}$$

and thus we have

$$\mathscr{M}^{a_\ell}_{\xi\mathscr{A}}(t,y) = \frac{g(t,y)}{\mathrm{Ai}'(a_\ell)^2} \int_0^\infty du\, e^{-ut/2}\mathrm{Ai}(a_\ell+u)\mathrm{Ai}(y+u), \quad \ell \in \mathbb{N}.$$

Then the correlation kernel is given by

$$\begin{aligned}&\sum_{\ell\in\mathbb{N}} p_{t^2/4}(s,x|0,a_\ell)\mathscr{M}^{a_\ell}_{\xi\mathscr{A}}(t,y) - \mathbf{1}_{(s>t)}p_{t^2/4}(s,x|t,y) \\ &= \frac{g(t,y)}{g(s,x)}\mathbb{K}_{\xi\mathscr{A}}(s,x;t,y)\end{aligned}$$

with

$$\mathbb{K}_{\xi^{\mathscr{A}}}(s,x;t,y) = \sum_{\ell\in\mathbb{N}} \frac{p_{\text{Ai}}(s,x|a_\ell)}{\text{Ai}'(a_\ell)^2} \int_0^\infty du\, e^{-ut/2}\text{Ai}(a_\ell+u)\text{Ai}(y+u)$$
$$-\mathbf{1}_{(s>t)} p_{\text{Ai}}(s-t,x|y),$$

where (3.173) and (3.174) were used.

Based on the above calculations, the following statement was proved in [83].

Theorem 3.7 *Under the initial configuration $\xi^{\mathscr{A}}$ given by all zeros of the Airy functions, the infinite system of SDEs*

$$dX_i(t) = dB_i^{x_i}(t) + \frac{t}{2}dt$$
$$+ \lim_{N\to\infty}\left(d_1 + \sum_{k=1}^{N}\frac{1}{a_k} + \sum_{\substack{1\le j\le N,\\ j\ne i}} \frac{1}{X_i(t)-X_j(t)}\right)dt, \quad (3.182)$$

$i\in\mathbb{N}$, $t\ge 0$, has a solution. The solution is given by a determinantal process $(\Xi, \mathbb{P}^{\xi^{\mathscr{A}}})$, whose correlation kernel is given by

$$\mathbb{K}_{\xi^{\mathscr{A}}}(s,x;t,y) = \sum_{\ell\in\mathbb{N}} p_{\text{Ai}}(s,x|a_\ell)\widetilde{\mathscr{M}}^{a_\ell}_{\xi^{\mathscr{A}}}(t,y) - \mathbf{1}_{(s>t)}p_{\text{Ai}}(s-t,x|y),$$
$$(s,x),(t,y)\in[0,\infty)\times\mathbb{R} \quad (3.183)$$

with

$$\widetilde{\mathscr{M}}^{a_\ell}_{\xi^{\mathscr{A}}}(t,y) = \frac{1}{\text{Ai}'(a_\ell)^2}\int_0^\infty du\, e^{-ut/2}\text{Ai}(a_\ell+u)\text{Ai}(y+u). \quad (3.184)$$

We call $(\Xi, \mathbb{P}^{\xi^{\mathscr{A}}})$ the *nonequilibrium Airy process*, because it exhibits the following relaxation phenomenon.

Proposition 3.9 *The infinite particle system $(\Xi, \mathbb{P}^{\xi^{\mathscr{A}}})$ starting from the zeros of the Airy function shows a relaxation phenomenon to the determinantal process $(\Xi, \mathbf{P}_{\text{Airy}})$, which is specified by the extended Airy kernel $\mathbf{K}_{\text{Airy}}$ obtained by the soft-edge scaling limit from the extended Hermite kernel.*

Proof By (3.171), (3.183) is written as

$$
\begin{aligned}
&\mathbb{K}_{\xi^{\mathscr{A}}}(s, x; t, y)\\
&= \int_0^\infty du \int_{\mathbb{R}} dw\, e^{-ut/2+ws/2} \mathrm{Ai}(y+u)\mathrm{Ai}(x+w) \sum_{\ell\in\mathbb{N}} \frac{\mathrm{Ai}(u+a_\ell)\mathrm{Ai}(w+a_\ell)}{\mathrm{Ai}'(a_\ell)^2}\\
&- \mathbf{1}_{(s>t)} p_{\mathrm{Ai}}(s-t, x|y).
\end{aligned}
$$

The functions $\{\mathrm{Ai}(x+a_\ell)/\mathrm{Ai}'(a_\ell)\}_{\ell\in\mathbb{N}}$ form an orthogonal basis for $f \in L^2(0,\infty)$ (see Sect. 4.12 in [135]) and the completeness is also established as

$$
\sum_{\ell\in\mathbb{N}} \frac{\mathrm{Ai}(x+a_\ell)\mathrm{Ai}(y+a_\ell)}{\mathrm{Ai}'(a_\ell)^2} dy = \delta_x(\{y\})dy, \quad x, y \in (0,\infty). \tag{3.185}
$$

Then we have the decomposition

$$
\mathbb{K}_{\xi^{\mathscr{A}}}(s, x; t, y) = \mathbf{K}_{\mathrm{Airy}}(s, x; t, y) + R_{\xi^{\mathscr{A}}}(s, x; t, y)
$$

with the extended Airy kernel $\mathbf{K}_{\mathrm{Airy}}$ given by (3.132) and

$$
\begin{aligned}
R_{\xi^{\mathscr{A}}}(s, x; t, y) = &\int_0^\infty du \int_{-\infty}^0 dw\, e^{-ut/2+ws/2} \mathrm{Ai}(u+y)\mathrm{Ai}(w+x)\\
&\times \sum_{\ell\in\mathbb{N}} \frac{\mathrm{Ai}(u+a_\ell)\mathrm{Ai}(w+a_\ell)}{\mathrm{Ai}'(a_\ell)^2}.
\end{aligned}
$$

Since for any fixed $s, t > 0$ $\lim_{T\to\infty} |R_{\xi^{\mathscr{A}}}(s+T, x; t+T, y)| \to 0$ uniformly on any compact subset of $\mathbb{R}^2$,

$$
\lim_{T\to\infty} \mathbb{K}_{\xi^{\mathscr{A}}}(s+T, x; t+T, y) = \mathbf{K}_{\mathrm{Airy}}(s, x; t, y)
$$

holds in the same sense. This completes the proof. □

A *Dirichlet form approach* has been developed to construct *equilibrium dynamics* of interacting infinite-particle systems [47, 109, 127, 132, 144]. Equilibrium determinantal processes including the sine process $(\Xi, \mathbf{P}_{\sin})$ and the Airy process $(\Xi, \mathbf{P}_{\mathrm{Airy}})$ with infinite numbers of particles are studied by Osada in [110–112].

3.11 Tracy–Widom Distribution

3.11.1 Distribution Function of the Maximum Position of Particles

Consider a determinantal point process (Ξ, P) with an infinite number of particles on $\mathbb{R}$ such that $\Xi = \sum_{i \in \mathbb{I}} \delta_{X_i} \in \mathfrak{M}_0$, where $\mathbb{I}$ denotes an infinite index set. Assume that the correlation kernel is given by $\mathrm{K}(x, y)$, $(x, y) \in \mathbb{R}^2$ [123, 126]. Two examples, $(\Xi, \mathrm{P}_{\sin})$ with the sine kernel $\mathrm{K}_{\sin}$ and $(\Xi, \mathrm{P}_{\mathrm{Airy}})$ with the Airy kernel $\mathrm{K}_{\mathrm{Airy}}$, were given in Sects. 3.9.2 and 3.9.3, respectively. With a test function $\chi \in \mathrm{C}_c(\mathbb{R})$ the generating function of spatial correlation functions defined by

$$\Psi[\chi] = \mathrm{E}\left[\prod_{i \in \mathbb{I}} \{1 + \chi(X_i)\}\right], \tag{3.186}$$

is expressed by a Fredholm determinant

$$\Psi[\chi] = \underset{(x,y)\in\mathbb{R}^2}{\mathrm{Det}} \Big[\delta_x(\{y\}) + \mathrm{K}(x, y)\chi(y)\Big]. \tag{3.187}$$

If we set $\chi(x) = -\mathbf{1}_{(x \geq s)}$ with a parameter $s \in \mathbb{R}$, (3.186) becomes

$$\begin{aligned}\Psi[-\mathbf{1}_{(\cdot \geq s)}] &= \mathrm{E}\left[\prod_{i \in \mathbb{I}} \{1 - \mathbf{1}_{(X_i \geq s)}\}\right] = \mathrm{E}\left[\prod_{i \in \mathbb{I}} \mathbf{1}_{(X_i < s)}\right] \\ &= \mathrm{P}\Big[X_i < s, {}^{\forall} i \in \mathbb{I}\Big] = \mathrm{P}\left[\max_{i \in \mathbb{I}} X_i < s\right].\end{aligned}$$

This is the distribution function of the *maximum position of particles*, and by (3.187), it has the Fredholm determinantal expression

$$\mathrm{P}\left[\max_{i \in \mathbb{I}} X_i < s\right] = \underset{(x,y)\in\mathbb{R}^2}{\mathrm{Det}} \Big[\delta_x(\{y\}) - \mathrm{K}_s(x, y)\Big], \tag{3.188}$$

where

$$\mathrm{K}_s(x, y) = \mathrm{K}(x, y)\mathbf{1}_{(y \geq s)}, \quad x, y, s \in \mathbb{R}. \tag{3.189}$$

For integrable functions $f_i(x, y)$, $i \in \mathbb{N}$, $(x, y) \in \mathbb{R}^2$, we use the following notations,

$$[f_1 f_2 \cdots f_n](x_1, x_{n+1}) = \int_{\mathbb{R}^{n-1}} f_1(x_1, x_2) f_2(x_2, x_3) \cdots f_n(x_n, x_{n+1}) dx_2 \ldots dx_n,$$

$n \in \{2, 3, \dots\}$, $x_1, x_{n+1} \in \mathbb{R}$. We regard $f_1 f_2 \cdots f_n$ as an operator such that its (x, y)-element is given by $[f_1 f_2 \cdots f_n](x, y)$, $(x, y) \in \mathbb{R}^2$. The trace of an operator f is defined by

$$\mathrm{Tr} f = \int_{\mathbb{R}} f(x, x) dx, \tag{3.190}$$

and if $\mathrm{Tr} f < \infty$, f is said to be a *trace class operator* [125]. Put $1(x, y) = \delta(x - y)$, $(x, y) \in \mathbb{R}^2$. The *resolvent* of K_a is defined by

$$\rho_a = \sum_{n=0}^{\infty} \mathrm{K}_a^n \equiv (1 - \mathrm{K}_a)^{-1}. \tag{3.191}$$

Let

$$\mathrm{R}_a \equiv \rho_a \mathrm{K}_a = \sum_{n=0}^{\infty} \mathrm{K}_a^{n+1}, \tag{3.192}$$

and

$$r(a) = \mathrm{R}_a(a, a) \equiv \lim_{y \to x} \mathrm{R}_a(x, y) \Big|_{x=a}, \quad a \in \mathbb{R}. \tag{3.193}$$

The correlation kernels K of determinantal point processes are trace class operators and the following exponential expression for (3.188) is proved (see, for instance, Lemmas 2.1 and 2.2 in [123]).

Lemma 3.10 *If r is integrable,*

$$\mathrm{P}\left[\max_{i \in \mathbb{I}} X_i < s\right] = \exp\left(-\int_s^{\infty} da\, r(a)\right), \quad s \in \mathbb{R}. \tag{3.194}$$

Proof The explicit expression of (3.188) is

$$\mathop{\mathrm{Det}}_{(x,y) \in \mathbb{R}^2} \Big[\delta_x(\{y\}) - \mathrm{K}_s(x, y)\Big] = 1 + \sum_{n=1}^{\infty} \frac{(-1)^n}{n!} I_n(s)$$

with

$$I_n(s) = \int_{\mathbb{R}^n} d\mathbf{x}_n \det_{1 \le i, j \le n} [\mathrm{K}_s(x_i, x_j)].$$

Using the Maclaurin expansion

$$\log(1-x) = -\sum_{n=1}^{\infty} \frac{x^n}{n}, \tag{3.195}$$

we can show that (Exercise 3.21)

$$\log\left(1 + \sum_{n=1}^{\infty} \frac{(-1)^n}{n!} I_n(s)\right) = -\sum_{n=1}^{\infty} \frac{1}{n} \mathrm{Tr}\, \mathrm{K}_s^n. \tag{3.196}$$

We rewrite this as

$$\begin{aligned}
\mathrm{Tr}\left(-\sum_{n=1}^{\infty} \frac{1}{n} \mathrm{K}_s^n\right) &= \int_s^{\infty} da \frac{\partial}{\partial a} \mathrm{Tr}\left(\sum_{n=1}^{\infty} \frac{1}{n} \mathrm{K}_a^n\right) \\
&= \int_s^{\infty} da\, \mathrm{Tr}\left(\sum_{n=1}^{\infty} \mathrm{K}_a^{n-1} \frac{\partial \mathrm{K}_a}{\partial a}\right) = \int_s^{\infty} da\, \mathrm{Tr}\left(\rho_a \frac{\partial \mathrm{K}_a}{\partial a}\right).
\end{aligned}$$

Here

$$\begin{aligned}
\frac{\partial \mathrm{K}_a}{\partial a}(x, y) &= \frac{\partial}{\partial a} \left\{\mathrm{K}(x, y) \mathbf{1}_{(y \geq a)}\right\} \\
&= \mathrm{K}(x, y) \frac{\partial}{\partial a} \mathbf{1}_{(y \geq a)} = -\mathrm{K}(x, y) \delta(y - a),
\end{aligned}$$

and hence

$$\begin{aligned}
\mathrm{Tr}\left(\rho_a \frac{\partial \mathrm{K}_a}{\partial a}\right) &= \int_{\mathbb{R}} dx \left[\rho_a \frac{\partial \mathrm{K}_a}{\partial a}\right](x, x) \\
&= \int_{\mathbb{R}} dx \int_{\mathbb{R}} dy\, \rho_a(x, y) \frac{\partial \mathrm{K}_a}{\partial a}(y, x) \\
&= -\int_{\mathbb{R}} dy\, \rho_a(a, y) \mathrm{K}(y, a) = -\mathrm{R}_a(a, a).
\end{aligned}$$

The proof is completed. □

For $(x, y) \in \mathbb{R}^2$, as $a \to \infty$, $\mathrm{K}_a(x, y) = \mathrm{K}(x, y)\mathbf{1}_{(y \geq a)} \to 0$. Then the definition (3.192) gives $\mathrm{R}_a(x, y) \simeq \mathrm{K}_a(x, y) = \mathrm{K}(x, y)\mathbf{1}_{(y \geq a)}$ in $a \to \infty$ for $(x, y) \in \mathbb{R}^2$. If we put $x = y = a$, we have

$$r(a) \simeq \mathrm{K}(a, a) \quad \text{as } a \to \infty. \tag{3.197}$$

Note that, by definitions (3.191) and (3.192), $\rho_a = 1 + \mathrm{R}_a$ and hence

$$\rho_a(x, y) = \delta(x - y) + \mathrm{R}_a(x, y), \quad (x, y) \in \mathbb{R}^2. \tag{3.198}$$

3.11.2 Integrals Involving Resolvent of Correlation Kernel

Let $N \in \mathbb{N}$, $t \in (0, \infty)$. Now we assume

$$\mathrm{K}(x, y) = \mathrm{K}^{(N,t)}_{\text{Hermite}}(x, y), \quad (x, y) \in \mathbb{R}^2. \tag{3.199}$$

The formula (3.113) is simply written as

$$\mathrm{K}(x, y) = \frac{A(x)B(y) - B(x)A(y)}{x - y}, \quad x \neq y, \tag{3.200}$$

with

$$A(x) = \left(\frac{N}{2}\right)^{1/4} \varphi_N\left(\frac{x}{\sqrt{2t}}\right), \quad B(x) = \left(\frac{N}{2}\right)^{1/4} \varphi_{N-1}\left(\frac{x}{\sqrt{2t}}\right). \tag{3.201}$$

We can prove the following (Exercise 3.22).

Lemma 3.11 *Let*

$$P_a(x) = \int_{\mathbb{R}} dz\, \rho_a(x, z)B(z) = [\rho_a B](x),$$
$$Q_a(x) = \int_{\mathbb{R}} dz\, \rho_a(x, z)A(z) = [\rho_a A](x). \tag{3.202}$$

Then

$$r(a) = \left[\frac{dQ_a(x)}{dx} P_a(x) - \frac{dP_a(x)}{dx} Q_a(x)\right]_{x=a}. \tag{3.203}$$

Using (3.201) and (3.202), we define the following integrals,

$$w(a) = \int_{\mathbb{R}} dx\, P_a(x)\mathbf{1}_{(x \geq a)} B(x) = \int_a^\infty dx\, P_a(x)B(x), \tag{3.204}$$
$$u(a) = \int_{\mathbb{R}} dx\, Q_a(x)\mathbf{1}_{(x \geq a)} A(x) = \int_a^\infty dx\, Q_a(x)A(x). \tag{3.205}$$

Then the following is proved (Exercise 3.23).

Lemma 3.12 *The following equations hold:*

$$\frac{dP_a(x)}{dx} = \frac{x}{2t}P_a(x) - \left(\sqrt{\frac{N}{t}} + \frac{w(a)}{t}\right)Q_a(x) + \mathrm{R}_a(x,a)P_a(a),$$
$$\frac{dQ_a(x)}{dx} = -\frac{x}{2t}Q_a(x) + \left(\sqrt{\frac{N}{t}} - \frac{u(a)}{t}\right)P_a(x) + \mathrm{R}_a(x,a)Q_a(a), \quad (3.206)$$

and

$$\mathrm{R}_a(x,x) = \left\{-\frac{x}{t}P_a(x)Q_a(x) + \left(\sqrt{\frac{N}{t}} - \frac{u(a)}{t}\right)P_a(x)^2 + \left(\sqrt{\frac{N}{t}} + \frac{w(a)}{t}\right)Q_a(x)^2 + \mathrm{R}_a(x,a)\{Q_a(a)P_a(x) - Q_a(x)P_a(a)\}\right\}\mathbf{1}_{(x\geq a)}. \quad (3.207)$$

3.11.3 Nonlinear Third-Order Differential Equation

Let

$$p(a) = P_a(a), \quad q(a) = Q_a(a), \quad a \in \mathbb{R}. \quad (3.208)$$

By the definition (3.193), (3.207) gives

$$r(a) = -\frac{a}{t}p(a)q(a) + \left(\sqrt{\frac{N}{t}} - \frac{u(a)}{t}\right)p(a)^2 + \left(\sqrt{\frac{N}{t}} + \frac{w(a)}{t}\right)q(a)^2. \quad (3.209)$$

Its derivative is

$$r'(a) = -\frac{1}{t}p(a)q(a) - \frac{a}{t}p'(a)q(a) - \frac{a}{t}p(a)q'(a) - \frac{u'(a)}{t}p(a)^2 + 2\left(\sqrt{\frac{N}{t}} - \frac{u(a)}{t}\right)p(a)p'(a) + \frac{w'(a)}{t}q(a)^2 + 2\left(\sqrt{\frac{N}{t}} + \frac{w(a)}{t}\right)q(a)q'(a). \quad (3.210)$$

We find the following system of differential equations (Exercise 3.24).

Lemma 3.13 *For* $a \in \mathbb{R}$,

$$p'(a) = \frac{a}{2t}p(a) - \left(\sqrt{\frac{N}{t}} + \frac{w(a)}{t}\right)q(a), \tag{3.211}$$

$$q'(a) = -\frac{a}{2t}q(a) + \left(\sqrt{\frac{N}{t}} - \frac{u(a)}{t}\right)p(a), \tag{3.212}$$

$$w'(a) = -p(a)^2, \tag{3.213}$$

$$u'(a) = -q(a)^2. \tag{3.214}$$

Inserting (3.211)–(3.214) into (3.210) gives a remarkably simple equation,

$$r'(a) = -\frac{1}{t}p(a)q(a). \tag{3.215}$$

Moreover, Tracy and Widom derived the following result [136, 137].

Proposition 3.10 *The function* $r(a)$ *solves the following nonlinear third-order differential equation,*

$$r'''(a) - \left(\frac{a^2}{t^2} - \frac{4N}{t}\right)r'(a) + \frac{a}{t^2}r(a) + 6r'(a)^2 = 0. \tag{3.216}$$

Proof By (3.211) and (3.212), we have

$$\begin{aligned}(p(a)q(a))' &= p'(a)q(a) + p(a)q'(a) \\ &= \left(\sqrt{\frac{N}{t}} - \frac{u(a)}{t}\right)p(a)^2 - \left(\sqrt{\frac{N}{t}} + \frac{w(a)}{t}\right)q(a)^2.\end{aligned} \tag{3.217}$$

On the other hand, (3.213) and (3.214) give

$$\begin{aligned}&\left(\sqrt{\frac{N}{t}}(u(a) - w(a)) + \frac{1}{t}u(a)w(a)\right)' \\ &= \sqrt{\frac{N}{t}}(u'(a) - w'(a)) + \frac{1}{t}(u'(a)w(a) + u(a)w'(a)) \\ &= \left(\sqrt{\frac{N}{t}} - \frac{u(a)}{t}\right)p(a)^2 - \left(\sqrt{\frac{N}{t}} + \frac{w(a)}{t}\right)q(a)^2.\end{aligned} \tag{3.218}$$

Then, we find the equality

$$(p(a)q(a))' = \left(\sqrt{\frac{N}{t}}(u(a) - w(a)) + \frac{1}{t}u(a)w(a)\right)'. \tag{3.219}$$

For finite x, $\mathbf{1}_{(x \geq a)} \to 0$ as $a \to \infty$, and $A(a) \to 0$, $B(a) \to 0$ as $a \to \infty$. Therefore $p(a)$, $q(a)$, $w(a)$ and $u(a)$ all become zero as $a \to \infty$. By integrating both sides of (3.219) from a to ∞, we obtain the equality

$$p(a)q(a) = \sqrt{\frac{N}{t}}(u(a) - w(a)) + \frac{1}{t}u(a)w(a). \tag{3.220}$$

If we use (3.217), the derivative of (3.215) is written as

$$r''(a) = -\frac{1}{t}\left\{\left(\sqrt{\frac{N}{t}} - \frac{u(a)}{t}\right)p(a)^2 - \left(\sqrt{\frac{N}{t}} + \frac{w(a)}{t}\right)q(a)^2\right\},$$

and then

$$\begin{aligned} r'''(a) = &-\frac{2}{t^2}p(a)^2q(a)^2 - \frac{a}{t^2}\left\{\left(\sqrt{\frac{N}{t}} - \frac{u(a)}{t}\right)p(a)^2 + \left(\sqrt{\frac{N}{t}} + \frac{w(a)}{t}\right)q(a)^2\right\} \\ &+ \frac{4}{t}p(a)q(a)\left[\frac{N}{t} - \frac{1}{t}\left\{\sqrt{\frac{N}{t}}(u(a) - w(a)) + \frac{1}{t}u(a)w(a)\right\}\right], \end{aligned} \tag{3.221}$$

where (3.211)–(3.214) were used. By (3.209) and (3.220), (3.221) is rewritten as

$$r'''(a) = -\frac{a}{t^2}r(a) - \left(\frac{a^2}{t^3} - \frac{4N}{t^2}\right)p(a)q(a) - \frac{6}{t^2}(p(a)q(a))^2.$$

By combining it with (3.215), we obtain (3.216). This completes the proof. □

3.11.4 *Soft-Edge Scaling Limit*

For $N \in \mathbb{N}$, $t \in (0, \infty)$, we perform the variable transformation $a \to u$ by

$$a = 2\sqrt{Nt} + \sqrt{t}N^{-1/6}u \iff u = (a - 2\sqrt{Nt})t^{-1/2}N^{1/6}. \tag{3.222}$$

Since $\partial/\partial a = t^{-1/2}N^{1/6}\partial/\partial u$, if we set $\widetilde{r}(u) = t^{1/2}N^{-1/6}r(a)$ with (3.222), (3.216) is transformed into

$$\widetilde{r}'''(u) - 4u\widetilde{r}'(u) + 2\widetilde{r}(u) + 6\widetilde{r}'(u)^2 - N^{-2/3}u\{u\widetilde{r}'(u) - \widetilde{r}(u)\} = 0.$$

On the other hand, (3.194) is written as $\exp(-\int_{(s-2\sqrt{Nt})t^{-1/2}N^{1/6}}^{\infty} du\, \widetilde{r}(u))$. Let $x = (s-2\sqrt{Nt})t^{-1/2}N^{1/6}$. Then

$$\max_{1\le i\le N} X_i(t) < s \iff \frac{\max_{1\le i\le N} X_i(t) - 2\sqrt{Nt}}{t^{1/2}N^{-1/6}} < x.$$

Therefore, we have the following limit,

$$\lim_{N\to\infty} \mathrm{P}_{\mathrm{Hermite}}^{(N,t)} \left[\frac{\max_{1\le i\le N} X_i(t) - 2\sqrt{Nt}}{t^{1/2}N^{-1/6}} < x \right] = \exp\left(-\int_x^{\infty} du\, \widetilde{r}(u)\right), \tag{3.223}$$

where $\widetilde{r}(u)$ solves the equation

$$\widetilde{r}'''(u) - 4u\widetilde{r}'(u) + 2\widetilde{r}(u) + 6\widetilde{r}'(u)^2 = 0. \tag{3.224}$$

With (3.222), the BM scaling variable $a/\sqrt{2t}$ behaves as

$$\frac{a}{\sqrt{2t}} = \sqrt{2N} + \frac{1}{\sqrt{2}} N^{-1/6} u,$$

which is the same as (3.133). Then the present limit $N \to \infty$ realizes the *soft-edge scaling limit* discussed in Sect. 3.9.3. By Proposition 3.7, we can conclude that the left-hand side of (3.223) is equal to [44]

$$\mathrm{P}_{\mathrm{Airy}}\left[\max_{i\in\mathbb{N}} X_i < x\right] = \operatorname*{Det}_{(u,v)\in\mathbb{R}^2}\left[\delta_u(\{v\}) - \mathrm{K}_{\mathrm{Airy}}(u,v)\mathbf{1}_{(v\ge x)}\right], \quad x\in\mathbb{R},$$

where the Airy kernel, $\mathrm{K}_{\mathrm{Airy}}$, is given by (3.136). Since $\mathrm{P}_{\mathrm{Airy}}$ is a stationary probability measure, this distribution obtained in the limit (3.223) does not depend on time $t \in (0,\infty)$.

3.11.5 Painlevé II and Limit Theorem of Tracy and Widom

Let

$$\widetilde{r}(u) = \int_u^{\infty} dv\, f(v)^2. \tag{3.225}$$

Then (3.224) is written as

$$f'(u)^2 + f(u)f''(u) - 2uf(u)^2 - \int_u^{\infty} dv\, f(v)^2 - 3f(u)^4 = 0.$$

If we differentiate this equation by u, we obtain

$$\left\{ f(u)\frac{d}{du} + 3f'(u) \right\} \left\{ f''(u) - uf(u) - 2f(u)^3 \right\} = 0.$$

Here we consider the equation

$$f''(u) = uf(u) + 2f(u)^3, \tag{3.226}$$

which is a special case of the *Painlevé II equation* (see, for instance, Chap. 21 and Appendix A.45 in [101], Chap. 8 in [45], Chap. 3 in [5], and Chap. 9 in [3]). Since (3.225) gives

$$\begin{aligned} \int_x^\infty du\, \widetilde{r}(u) &= \int_x^\infty du \int_u^\infty dv\, f(v)^2 \\ &= \int_x^\infty dv\, f(v)^2 \int_x^v du = \int_x^\infty dv\, (v-x) f(v)^2, \end{aligned}$$

the RHS of (3.223) is written as $\exp(-\int_x^\infty dv\, (v-x) f(v)^2)$.

By (3.197), in the soft-scaling limit, we find

$$\widetilde{r}(u) \simeq \mathrm{K}_{\mathrm{Airy}}(u,u) \quad \text{as } u \to \infty.$$

We note that the integral representation of $\mathrm{K}_{\mathrm{Airy}}$ (3.135) gives

$$\mathrm{K}_{\mathrm{Airy}}(u,u) = \int_0^\infty dw\, \mathrm{Ai}(u+w)^2 = \int_u^\infty dv\, \mathrm{Ai}(v)^2.$$

Comparing this with (3.225), we can conclude that

$$f(u) \simeq \mathrm{Ai}(u) \quad \text{as } u \to \infty. \tag{3.227}$$

Hastings and McLeod [56] proved that the Painlevé II equation (3.226) has a unique solution $f_{\mathrm{HM}}(u)$, which satisfies (3.227).

Now we arrive at the following limit theorem for the maximum position of particles of the Dyson model with an infinite number of particles.

Theorem 3.8 *For any $t \in (0,\infty)$, the probability*

$$\lim_{N\to\infty} \mathrm{P}_{\mathrm{Hermite}}^{(N,t)} \left[\frac{\max_{1\le i\le N} X_i(t) - 2\sqrt{Nt}}{t^{1/2}N^{-1/6}} < x \right] = \mathrm{P}_{\mathrm{Airy}} \left[\max_{i\in\mathbb{N}} X_i < x \right], \quad x \in \mathbb{R}, \tag{3.228}$$

has the following two expressions,

$$F_{\mathrm{TW}}(x) = \operatorname*{Det}_{(u,v)\in\mathbb{R}^2}\left[\delta_u(\{v\}) - \mathbf{K}_{\mathrm{Airy}}(u,v)\mathbf{1}_{(v\geq x)}\right] \tag{3.229}$$

$$= \exp\left(-\int_x^{\infty} dv\,(v-x) f_{HM}(v)^2\right), \quad x \in \mathbb{R}. \tag{3.230}$$

Here the former is the Fredholm determinantal expression, and the latter is the expression in terms of the Hastings–McLeod solution f_{HM} *of the Painlevé II equation (3.226).*

The probability distribution function (3.230) is called the *Tracy–Widom distribution* [136, 137]. It has the probability density function

$$p_{\mathrm{TW}}(x) = \frac{dF_{\mathrm{TW}}(x)}{dx}, \quad x \in \mathbb{R}. \tag{3.231}$$

Numerical values of the mean, variance, skewness, and kurtosis are the following (see [138], Sect. 9.4.2 in [45], and [115]),

$$\mu_{\mathrm{TW}} = \int_{-\infty}^{\infty} x p_{\mathrm{TW}}(x)dx = -1.771086807,$$

$$\sigma_{\mathrm{TW}}^2 = \int_{-\infty}^{\infty} (x-\mu_{\mathrm{TW}})^2 p_{\mathrm{TW}}(x)dx = 0.813194792,$$

$$S_{\mathrm{TW}} = \int_{-\infty}^{\infty} \left(\frac{x-\mu_{\mathrm{TW}}}{\sigma_{\mathrm{TW}}}\right)^3 p_{\mathrm{TW}}(x)dx = 0.224084203,$$

$$K_{\mathrm{TW}} = \int_{-\infty}^{\infty} \left(\frac{x-\mu_{\mathrm{TW}}}{\sigma_{\mathrm{TW}}}\right)^4 p_{\mathrm{TW}}(x)dx - 3 = 0.093448087. \tag{3.232}$$

Figure 3.2 shows the comparison between $p_{\mathrm{TW}}(x)$ and the probability density function of the Gaussian distribution

$$p_{\mathrm{G}}(x) = \frac{1}{\sqrt{2\pi\sigma^2}} e^{-(x-\mu)^2/2\sigma^2}, \quad x \in \mathbb{R}, \tag{3.233}$$

with the same values of mean and variance as the Tracy–Widom distribution given by (3.232) ($\mu = \mu_{\mathrm{TW}}, \sigma^2 = \sigma_{\mathrm{TW}}^2$). The difference between p_{TW} and p_{G} can be shown better, if we represent them in the semi-log plots as given by Fig. 3.3.

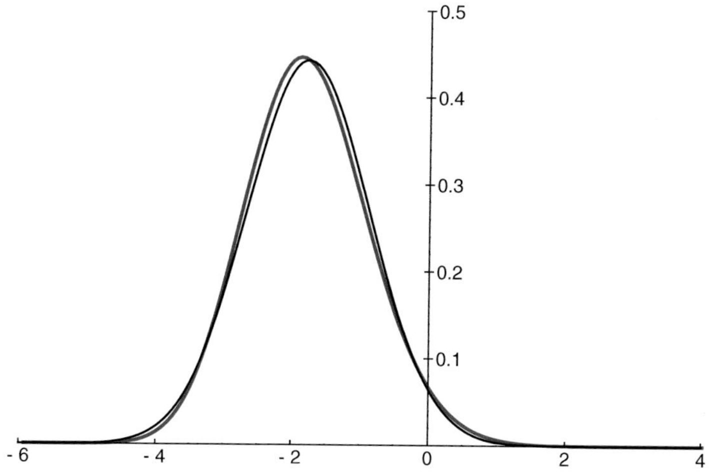

Fig. 3.2 The probability density function of the Tracy–Widom distribution (3.231) is shown by a *red curve*. The *black curve* shows the probability density function of the Gaussian distribution (3.233) with the same values of mean and variance as the Tracy–Widom distribution given by (3.232) ($\mu = \mu_{\text{TW}}, \sigma^2 = \sigma_{\text{TW}}^2$)

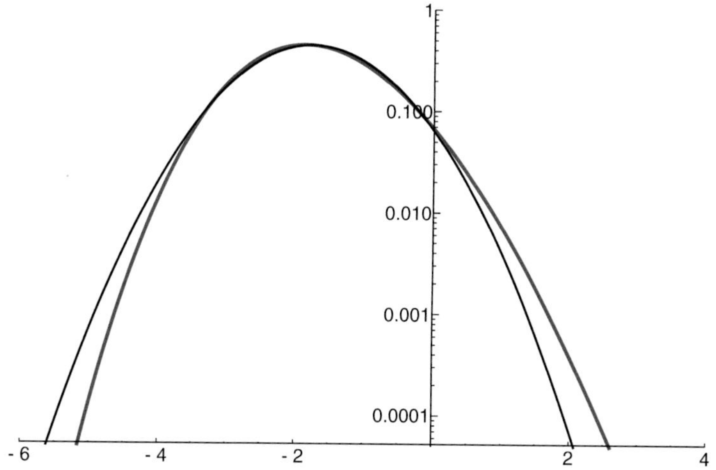

Fig. 3.3 The semi-log plots of the probability density function of the Tracy–Widom distribution (3.231) (*the red curve*) and that of the Gaussian distribution (3.233) with the same values of mean and variance (*the black curve*)

3.12 Beyond Determinantal Processes

In this last chapter, we have discussed the interacting particle system (3.35), which is the special case where $\beta = 2$ for Dyson's Brownian motion model (3.3). Only with this special choice of parameter, **[Aspect 1]** (see Sect. 3.2) and **[Aspect 2]** (see Sect. 3.3) are established and the interacting particle system (3.35) can be constructed as a multivariate extension of the one-dimensional diffusion process $\text{BES}^{(3)}$. In particular, **[Aspect 2]** makes the Dyson model (3.35) a *solvable model* in the sense that any generating function of correlation functions is given by a Fredholm determinant, and hence all spatio-temporal correlation functions are expressed by determinants which are controlled by a single continuous function called the correlation kernel. We have called interacting particle systems having such strong *solvability* determinantal processes [82, 84]. In order to link **[Aspect 2]** and the solvability, we have introduced the notion of determinantal martingale representation (DMR) and proved useful formulas, (3.88) with (3.89) for the initial configuration $\xi \in \mathfrak{M}_0$ and (3.104) for $\xi \in \mathfrak{M} \setminus \mathfrak{M}_0$, which give the correlation kernels $\mathbb{K}_\xi$, using the martingale functions $\{\mathscr{M}_\xi^v : v \in \text{supp } \xi\}$. One of the highlights of determinantal structures is the appearance of the Tracy–Widom distribution, which has the Fredholm determinantal expression (3.229) as well as the expression (3.230), using the Hastings–McLeod solution of the Painlevé II equation.

There are many directions in recent developments of the theory. The following are some of them.

(1) We have assumed that the initial configuration $\xi = \sum_{i=1}^N \delta_{x_i} \in \mathfrak{M}$ is deterministic. For $\sigma \in \mathbb{R}$, $\beta \geq 1$, let

$$\mu_{N,\sigma^2}^{(\beta)}(\xi) = \frac{\sigma^{-N\{\beta(N-1)+2)/2}}{C_N^{(\beta)}} e^{-|\mathbf{x}|^2/2\sigma^2} |h_N(\mathbf{x})|^\beta$$

with

$$C_N^{(\beta)} = \frac{(2\pi)^{N/2}}{N!} \prod_{i=1}^N \frac{\Gamma(i\beta/2+1)}{\Gamma(\beta/2+1)},$$

and $|\mathbf{x}|^2 = \sum_{i=1}^N x_i^2$. If ξ is distributed with the probability density $\mu_{N,\sigma^2}^{(2)}$ which has unitary symmetry with variance σ^2, the determinantal structure of multitime correlation functions is maintained but the correlation kernel is replaced by the time shift $t \mapsto t + \sigma^2$ of the correlation kernel for the special initial configuration $\xi = N\delta_0$ [76]. In [69], it was shown that if the distribution of ξ has the probability density $\mu_{N,\sigma^2}^{(1)}$ with orthogonal symmetry, the system becomes a *Pfaffian process*, in which all spatio-temporal correlation functions are given by Pfaffians. Here the *Pfaffian* is defined for a skew-symmetric $2N \times 2N$ matrix $A = (a_{ij})$, $N \in \mathbb{N}$ as

$$\text{Pf}(A) = \frac{1}{N!} \sum_{\sigma}' \text{sgn}(\sigma) a_{\sigma(1)\sigma(2)} a_{\sigma(3)\sigma(4)} \cdots a_{\sigma(2N-1)\sigma(2N)},$$

where the summation $\sum_{\sigma}'$ is extended over all permutations $\sigma \in \mathscr{S}_{2N}$ with the restriction $\sigma(2k-1) < \sigma(2k), k = 1, 2, \dots, N$. See [20, 46, 77, 81, 98, 103, 104, 106] for Pfaffian processes.
(2) Even though the system is not determinantal, if we properly choose the initial condition, then some observables can have useful Fredholm determinantal expressions. An important example is found in the *asymmetric simple exclusion process* (ASEP) defined on the one-dimensional integer lattice $\mathbb{Z}$. It is *not* determinantal, but Tracy and Widom [141] proved that in the case of the step initial configuration, $\xi = \sum_{n \in \mathbb{N}} \delta_n$, the probability distribution function of particle positions is given by an integral whose integrand involves a Fredholm determinant. Using this result as a starting point, Sasamoto and Spohn [119, 120] and Amir et al. [4] obtained an exact solution of the stochastic partial differential equation introduced by Kardar et al. [66] to describe growing interfaces in 1+1 dimensions,

$$\frac{\partial h(t,x)}{\partial t} = \nu \frac{\partial^2 h(t,x)}{\partial x^2} + \frac{\lambda}{2}\left(\frac{\partial h(t,x)}{\partial x}\right)^2 + \sqrt{D}\dot{W}(t,x), \tag{3.234}$$

where $h(t,x)$ is the height profile at time $t \geq 0$ and $x \in \mathbb{R}$, ν, λ, D are positive parameters, and $\{\dot{W}(t,x)\}_{t \geq 0, x \in \mathbb{R}}$ denotes the space-time *Gaussian white noise*, having the covariance

$$\mathrm{E}[\dot{W}(s,x)\dot{W}(t,y)] = \delta(s-t)\delta(x-y), \quad (s,x),(t,y) \in [0,\infty) \times \mathbb{R}.$$

The equation (3.234) is called the *Kardar–Parisi–Zhang* (KPZ) *equation*. The mathematical justification of such a nonlinear stochastic differential equation itself is an important research subject [13, 19, 49, 55]. The solution obtained by [4, 119, 120] shows that the fluctuation of the KPZ interface exhibits a crossover from the Gaussian distribution at short time to the Tracy–Widom distribution at long times. The universality of Tracy–Widom distributions was also clarified by the replica method with the Bethe ansatz [25, 35, 36].
(3) Besides **[Aspect 1]** and **[Aspect 2]**, the following aspect of BES$^{(3)}$ is known as *Pitman's theorem* [113]. Let $(R^0(t))_{t \geq 0}$ be a BES$^{(3)}$ started at 0, $(W(t))_{t \geq 0}$ be a BM started at 0, and $M(t) = \max_{0 \leq s \leq t} W(s), \ t \geq 0$. Then

$$(R^0(t))_{t \geq 0} \stackrel{\text{(law)}}{=} (2M(t) - W(t))_{t \geq 0}. \tag{3.235}$$

Let $X_N(t), \ t \geq 0$ denotes the rightmost particle of the Dyson model started at $\xi = N\delta_0$. As a multivariate extension of (3.235), the following equality is established,

$$(X_N(t))_{t \geq 0} \stackrel{\text{(law)}}{=} \left(\max_{\Delta_N([0,t])} \sum_{i=1}^{N} \{W_i(t_i) - W_i(t_{i-1})\} \right)_{t \geq 0}, \tag{3.236}$$

where $\{W_i(t)\}_{i=1}^N$, $t \geq 0$ are independent BMs started at 0, and the maximum is taken over all subdivisions $\Delta_N([0,t])$ of $[0,t]$ with $0 \equiv t_0 < t_1 < \cdots < t_{N-1} < t_N \equiv t$. This equality was first proved for each fixed time $t \geq 0$ by Gravner et al. [54] and Baryshnikov [12], and then the equality in probability law at the level of processes was proved by Bougerol and Jeulin [21] and O'Connell and Yor [108] (see also [14, 15, 29]). Moreover, Warren [143] gave a new construction of the Dyson model started at $\xi = N\delta_0$, $(\Xi, \mathbb{P}^{N\delta_0})$, which can be regarded as a generalization of (3.236). Matsumoto and Yor generalized Pitman's theorem by considering exponential functionals of BM [99, 100] and its multivariate version was introduced by O'Connell [107]. The *O'Connell process* is a softened version (a geometric lifting) with a parameter $a > 0$ of the Dyson model (i.e. the noncolliding Brownian motion), $\Xi(t) = \sum_{i=1}^N \delta_{X_i(t)}$, $t \geq 0$, such that neighboring particles can change the order of positions in one dimension within the characteristic length a. Construction of the O'Connell process as a system of conditional Brownian motions was given in [68, 70, 71]. This process, here denoted by $\Xi^a = \sum_{i=1}^N \delta_{X_i^a(t)}$, $t \geq 0$, is not determinantal. Under a special entrance law, however, Borodin and Corwin [19] gave a Fredholm determinant expression for the expectation of an observable,

$$\Theta^a(X_N^a(t) - x) \equiv \exp\left[-e^{-\{X_N^a(t)-x\}/a}\right], \quad x \in \mathbb{R},$$

which is a smoothing of the indicator function $\mathbf{1}_{(X_N(t)>x)}$, $x \in \mathbb{R}$ of the rightmost particle of the Dyson model. For this special observable, a DMR is given in [72]. Beyond determinantal processes, Borodin and Corwin proposed a general family of interacting particle systems called the *Macdonald process* [19]. This family has a hierarchical structure and the Dyson model and the O'Connell process (which is also called the Whittaker process) are located at the lowest and the second lowest level, respectively.

Exercises

3.1 Prove that the SDE of the relative coordinate defined by $X_{\mathrm{r}}(t) = \{X_2(t) - X_1(t)\}/\sqrt{2}$ for the two-particle system (3.1) is given by (3.2).

3.2 (i) Prove (3.14) for the Hermitian-matrix-valued Brownian motion $\mathsf{H}(t) = \mathsf{B}^{\mathbf{x}}(t)$, $t \geq 0$ given by (3.4).
(ii) Show that $\Gamma_{ij}(t) = 1$, $1 \leq i, j \leq N$ in this case.

3.3 (i) Prove that $q_N(t, \mathbf{y}|\mathbf{x})$ satisfies the N-dimensional diffusion equation

$$\frac{\partial}{\partial t} q_N(t, \mathbf{y}|\mathbf{x}) = \frac{1}{2}\Delta^{(N)} q_N(t, \mathbf{y}|\mathbf{x}). \tag{3.237}$$

(ii) Show the following equalities:

$$\frac{\partial}{\partial x_i}\frac{1}{h_N(\mathbf{x})} = -\frac{1}{h_N(\mathbf{x})}\sum_{\substack{1\le j\le N,\\ j\ne i}}\frac{1}{x_i - x_j}, \quad 1 \le i \le N. \tag{3.238}$$

(iii) Prove

$$\sum_{\substack{1\le i,j,k\le N,\\ i\ne j, i\ne k, j\ne k}}\frac{1}{(x_i - x_j)(x_i - x_k)} = 0. \tag{3.239}$$

(iv) Show that (3.32) satisfies the PDE (3.34) with the initial condition $p_N(0, \mathbf{y}|\mathbf{x}) = \delta(\mathbf{x} - \mathbf{y})$ for $\mathbf{x}, \mathbf{y} \in \mathbb{W}_N^{\mathrm{A}}$.

3.4 For an $N \times N$ matrix $A = (a_{ij})_{1\le i,j\le N}$, the determinant $\det A$ is defined by

$$\det A = \det_{1\le i,j\le N}[a_{ij}] = \sum_{\sigma\in\mathscr{S}_N}\mathrm{sgn}(\sigma)\prod_{i=1}^{N} a_{i\sigma(i)}. \tag{3.240}$$

(i) Prove the following:

$$\det_{1\le i,j\le N}[a_{ij}b_j] = \det_{1\le i,j\le N}[b_i a_{ij}] = \prod_{k=1}^{N} b_k \times \det_{1\le i,j\le N}[a_{ij}]. \tag{3.241}$$

This property is called the *multilinearity of determinants*.
(ii) Prove the equality (3.38).

3.5 Let $V_N(\mathbf{x}) = \det_{1\le i,j\le N}[x_j^{N-i}]$. Prove the equality

$$V_N(\mathbf{x}) = \prod_{1\le i<j\le N}(x_i - x_j). \tag{3.242}$$

3.6 For each partition $\mu = (\mu_1, \mu_2, \dots)$, let $|\mu| = \sum_{i\ge 1}\mu_i$. Prove that $s_\mu(\mathbf{x})$ defined by (3.40) is a homogeneous polynomial of $x_1, \dots, x_N$ of degree $|\mu|$.

3.7 Write down the Schur polynomials $s_\mu(\mathbf{x})$ for the following cases:
(i) $N = 2, \mu = (1)$. (ii) $N = 3, \mu = (2, 1)$.

3.8 Prove the equality (3.45).

3.9 For $n \in \mathbb{N}_0$, assume that $M_n(x)$ is a polynomial of degree n, $M_n(x) = \sum_{i=0}^{n} b_{n,i}x^i$, $b_{n,n} \ne 0$. Then prove the equality $\det_{1\le i,j\le N}[M_{j-1}(x_i)] = \prod_{i=1}^{N} b_{i-1,i-1} \times h_N(\mathbf{x})$. In particular, if $\{M_n\}_{n\in\mathbb{N}_0}$ are monic, *i.e.*, $b_{n,n} = 1, {}^\forall n \in \mathbb{N}_0$, then $\det_{1\le i,j\le N}[M_{j-1}(x_i)] = h_N(\mathbf{x})$.

3.10 Remember that $\mathscr{G}_\xi$ is defined by (3.72), where p is the transition probability density (1.1) of BM.

(i) Show the equality

$$\int_{\mathbb{R}^2} \xi^{\otimes 2}(d\mathbf{v}) \times \mathrm{E}^{(v_1,v_2)}\left[\chi_1(B_1(t_1))\chi_2(B_2(t_2)) \det \begin{pmatrix} \mathscr{M}_\xi^{v_1}(t_1, B_1(t_1)) & \mathscr{M}_\xi^{v_1}(t_2, B_2(t_2)) \\ \mathscr{M}_\xi^{v_2}(t_1, B_1(t_1)) & \mathscr{M}_\xi^{v_2}(t_2, B_2(t_2)) \end{pmatrix}\right]$$
$$= \int_{\mathbb{R}^2} dx_1 dx_2\, \chi_1(x_1)\chi_2(x_2) \det \begin{pmatrix} \mathscr{G}_\xi(t_1, x_1; t_1, x_1) & \mathscr{G}_\xi(t_1, x_1; t_2, x_2) \\ \mathscr{G}_\xi(t_2, x_2; t_1, x_1) & \mathscr{G}_\xi(t_2, x_2; t_2, x_2) \end{pmatrix}. \quad (3.243)$$

(ii) Show the equality

$$\int_{\mathbb{R}} \xi(dv)\mathrm{E}^v[\chi_1(B(t_1))\chi_2(B(t_2))\mathscr{M}_\xi^v(t_2, B(t_2))]$$
$$= \int_{\mathbb{R}^2} dx_1 dx_2\, \chi_1(x_1)\chi_2(x_2)\mathscr{G}_\xi(t_1, x_1; t_2, x_2)p(t_2 - t_1, x_2|x_1). \quad (3.244)$$

3.11 Let $\mathsf{I} = (\delta_{ij})_{i,j\in\mathbb{I}_N}$ be the $N \times N$ unit matrix. For an $N \times N$ matrix $\mathsf{M} = (m_{ij})_{i,j\in\mathbb{I}_N}$, the *characteristic polynomial* is defined by

$$f_{\mathsf{M}}(x) = \det(x\mathsf{I} + \mathsf{M}) = \det_{1\le i,j\le N}[x\delta_{ij} + m_{ij}], \quad x \in \mathbb{R}.$$

Prove the equality

$$f_{\mathsf{M}}(x) = \sum_{N'=0}^{N} x^{N-N'} \sum_{\mathbb{J}\subset\mathbb{I}_N, \sharp\mathbb{J}=N'} \det_{i,j\in\mathbb{J}}[m_{ij}]. \quad (3.245)$$

In particular, if we set $x = 1$ and $m_{ij} = K_{ij}\widehat{\chi}_j$, $1 \le i, j \le N$, we obtain the Fredholm expansion formula

$$\det_{1\le i,j\le N}[\delta_{ij} + K_{ij}\widehat{\chi}_j] = \sum_{N'=0}^{N} \sum_{\mathbb{J}\subset\mathbb{I}_N, \sharp\mathbb{J}=N'} \prod_{k\in\mathbb{J}} \widehat{\chi}_k \det_{i,j\in\mathbb{J}}[K_{ij}].$$

3.12 (i) Let $q_N^{(b)}(t-s, \mathbf{y}|\mathbf{x}) = \det_{1\le i,j\le N}[p^{(b)}(t-s, y_i|x_j)]$, $\mathbf{x}, \mathbf{y} \in \mathbb{W}_N^{\mathrm{A}}$, $0 \le s \le t$, where $p^{(b)}$ is given by (1.29). Prove that

$$-\frac{\partial}{\partial s}q_N^{(b)}(t-s, \mathbf{y}|\mathbf{x}) = \frac{1}{2}\Delta^{(N)}q_N^{(b)}(t-s, \mathbf{y}|\mathbf{x}) + b\sum_{i=1}^{N}\frac{\partial}{\partial x_i}q_N^{(b)}(t-s, \mathbf{y}|\mathbf{x}). \quad (3.246)$$

(ii) The backward Kolmogorov equation for the drifted Dyson model (3.90) is given by

$$-\frac{\partial}{\partial s} p_N^{(b)}(t-s, \mathbf{y}|\mathbf{x}) = \frac{1}{2}\Delta^{(N)} p_N^{(b)}(t-s, \mathbf{y}|\mathbf{x}) + b\sum_{i=1}^{N} \frac{\partial}{\partial x_i} p_N^{(b)}(t-s, \mathbf{y}|\mathbf{x})$$
$$+ \sum_{\substack{1\le i,j\le N,\\ i\ne j}} \frac{1}{x_i - x_j}\frac{\partial}{\partial x_i} p_N^{(b)}(t-s, \mathbf{y}|\mathbf{x}), \quad \mathbf{x}, \mathbf{y} \in \mathbb{W}_N^{\mathrm{A}},\ 0 \le s < t. \qquad (3.247)$$

Assume that $p_N^{(b)}(t-s, \mathbf{y}|\mathbf{x}) = c(\mathbf{y}) f(\mathbf{x}) q_N^{(b)}(t-s, \mathbf{y}|\mathbf{x})$ with a differentiable function f of $\mathbf{x}$. Derive the equation for $f(\mathbf{x})$.
(iii) Show that $f(\mathbf{x}) = 1/h_N(\mathbf{x})$ satisfies the above equation.
(iv) Prove that

$$p_N^{(b)}(t-s, \mathbf{y}|\mathbf{x}) = \frac{h_N(\mathbf{y})}{h_N(\mathbf{x})} q_N^{(b)}(t-s, \mathbf{y}|\mathbf{x}) \qquad (3.248)$$

is the unique solution of (3.247) satisfying the condition $\lim_{s\uparrow t} p_N^{(b)}(t-s, \mathbf{y}|\mathbf{x}) = \delta(\mathbf{y}-\mathbf{x})$.
(v) Show that (3.248) is equal to (3.92).

3.13 Prove the equality (3.98).

3.14 Show that, if $\xi \in \mathfrak{M}_0$, that is, ξ has no multiple points, (3.104) is equal to (3.88) with (3.89).

3.15 Let

$$S(x, y; a) = \sum_{n=0}^{\infty} \frac{H_n(x)H_n(y)}{2^n n!} a^n, \quad |a| < 1, x, y \in \mathbb{R}. \qquad (3.249)$$

(i) Show that S satisfies the following equations:

$$\frac{\partial S}{\partial x} = 2ayS - a\frac{\partial S}{\partial y}, \qquad (3.250)$$
$$\frac{\partial S}{\partial y} = 2axS - a\frac{\partial S}{\partial x}. \qquad (3.251)$$

(ii) From (3.250) and (3.251), derive the equation

$$\frac{\partial \log S}{\partial x} = \frac{2ya - 2xa^2}{1-a^2}. \qquad (3.252)$$

(iii) Show that the solution of (3.250) and (3.251) is given in the form

$$S(x, y; a) = c(a)e^{[2xya-(x^2+y^2)a^2]/(1-a^2)} \tag{3.253}$$

and determine $c(a)$.
(iv) Prove (3.110).

3.16 Following the instructions below, prove the equality

$$\sum_{n=0}^{N-1} \varphi_n(x)\varphi_n(y) = \sqrt{\frac{N}{2}} \frac{\varphi_N(x)\varphi_{N-1}(y) - \varphi_{N-1}(x)\varphi_N(y)}{x - y} \tag{3.254}$$

for $x \neq y$ by mathematical induction with respect to $N \in \mathbb{N}$. Equation (3.254) is called the *Christoffel–Darboux formula* for the Hermite orthonormal functions $\{\varphi_n(x)\}_{n\in\mathbb{N}_0}$.
(i) Prove (3.254) for $N = 1$.
(ii) Assume that (3.254) holds for a given $N \in \mathbb{N}$. Then show

$$\begin{aligned}&\text{(RHS) of (3.254)}\\ &= -\varphi_N(x)\varphi_N(y) + \sqrt{\frac{N+1}{2}} \frac{\varphi_{N+1}(x)\varphi_N(y) - \varphi_N(x)\varphi_{N+1}(y)}{x - y}.\end{aligned} \tag{3.255}$$

This implies that (3.254) holds even if we replace N by $N + 1$.

3.17 (i) Assume that A and B are differentiable functions and

$$F(x, y) = \frac{A(x)B(y) - B(x)A(y)}{x - y}, \quad x \neq y. \tag{3.256}$$

Show that

$$F(x, x) = \lim_{y\to x} F(x, y) = A'(x)B(x) - B'(x)A(x). \tag{3.257}$$

(ii) Derive (3.114) from (3.113) by taking the limit $y \to x$.

3.18 (i) Derive (3.136) from (3.135).
(ii) Derive (3.137) by taking the limit $y \to x$ in (3.136).

3.19 Using the expression (3.136) for (3.135) with $x \neq y$, derive the expression (3.180).

3.20 Prove (3.181) by following the instructions below.

(i) Show

$$\mathscr{I}[\mathrm{Ai}(W+u)|(t, y-t^2/4)] = e^{(y-t^2/4)^2/2t}\frac{1}{2\pi}\int_C dk\, e^{k^3/3-uk}$$
$$\times\frac{1}{\sqrt{2\pi t}}\int_{\mathbb{R}} dw\,\exp\left[-\frac{w^2}{2t}-\frac{\sqrt{-1}}{t}\left(y-\frac{t^2}{4}\right)w-\sqrt{-1}kw\right],$$

where C is the same contour as that for (3.158).
(ii) Perform the integral over w to give

$$\mathscr{I}[\mathrm{Ai}(W+u)|(t, y-t^2/4)] = \frac{1}{2\pi}\int_C dk\, e^{f(k)} \tag{3.258}$$

with

$$f(k) = \frac{k^3}{3}-\frac{t}{2}k^2-\left(y-\frac{t^2}{4}-u\right)k. \tag{3.259}$$

(iii) Complete a cube in (3.259) to prove (3.181).

3.21 Show that

$$\log\left(1+\sum_{n=1}^{3}\frac{(-1)^n}{n!}I_n(s)\right) = -\sum_{n=1}^{3}\frac{1}{n}\mathrm{Tr}\,\mathrm{K}_s^n + (\text{correction terms}), \tag{3.260}$$

$s \in \mathbb{R}$.

3.22 Prove Lemma 3.11 following the instructions below.
(i) Let M be an operator which multiplies a variable, $\mathrm{M}(x, y) = x\delta(x-y)$, $x, y \in \mathbb{R}$. Then $[\mathrm{M}f](x, y) = \int_{\mathbb{R}} x\delta(x-z)f(z, y)dz = xf(x, y)$, and $[fM](x, y) = f(x, y)y$. We introduce the commutator $[U, V] = UV - VU$. Using (3.198), derive the equality

$$[\mathrm{M}, \rho_a](x, y) = (x-y)\mathrm{R}_a(x, y). \tag{3.261}$$

(ii) If we use the definition of the resolvent ρ_a given by (3.191),

$$[\mathrm{M}, \rho_a] = \mathrm{M}(1-\mathrm{K}_a)^{-1} - (1-\mathrm{K}_a)^{-1}\mathrm{M}. \tag{3.262}$$

Derive the equality

$$[\mathrm{M}, \rho_a] = \rho_a[\mathrm{M}, \mathrm{K}_a]\rho_a. \tag{3.263}$$

(iii) Using (3.200), derive the expression

$$[\mathrm{M}, \rho_a](x, y) = Q_a(x) \int_{\mathbb{R}} dz\, B(z)\mathbf{1}_{(z \geq a)} \rho_a(z, y) \\ - P_a(x) \int_{\mathbb{R}} dz\, A(z)\mathbf{1}_{(z \geq a)} \rho_a(z, y). \tag{3.264}$$

(iv) Using the symmetry $\mathrm{K}(x, y) = \mathrm{K}(y, x)$, prove the equalities

$$\int_{\mathbb{R}} dz\, B(z)\mathbf{1}_{(z \geq a)} \rho_a(z, y) = P_a(y)\mathbf{1}_{(y \geq a)}, \\ \int_{\mathbb{R}} dz\, A(z)\mathbf{1}_{(z \geq a)} \rho_a(z, y) = Q_a(y)\mathbf{1}_{(y \geq a)}. \tag{3.265}$$

(v) Combination of (3.264) with (3.265) gives the equation

$$[\mathrm{M}, \rho_a](x, y) = \{Q_a(x)P_a(y) - P_a(x)Q_a(y)\}\mathbf{1}_{(y \geq a)}.$$

By identifying the above with (3.261), prove (3.203).

3.23 Prove Lemma 3.12 following the instructions below.
(i) We introduce the differential operator $\mathrm{D}(x, y) = \delta(x-y)\partial/\partial y$, $x, y \in \mathbb{R}$. Assume that $f(x, z) \to 0$, $g(z, y) \to 0$ as $z \to \pm\infty$. Show

$$[f\mathrm{D}](x, y) = -\frac{\partial f(x, y)}{\partial y}. \tag{3.266}$$

(ii) Using (3.200), show that $[\mathrm{D}, \mathrm{K}_a](x, y)$ is equal to

$$\frac{A'(x)B(y) - B'(x)A(y) + A(x)B'(y) - B(x)A'(y)}{x - y}\mathbf{1}_{(y \geq a)} + \mathrm{K}(x, y)\delta(y - a). \tag{3.267}$$

(iii) Using the recurrence relations for the Hermite orthonormal functions (3.108) obtained from (1.121) and (1.122), derive the equation

$$[\mathrm{D}, \mathrm{K}_a](x, y) = -\frac{1}{2t}\{A(x)B(y) + B(x)A(y)\}\mathbf{1}_{(y \geq a)} + \mathrm{K}(x, y)\delta(y - a). \tag{3.268}$$

(iv) Using (3.268), derive the equation

$$[\mathrm{D}, \rho_a](x, y) = -\frac{1}{2t}\{Q_a(x)P_a(y) + P_a(x)Q_a(y)\}\mathbf{1}_{(y \geq a)} + \mathrm{R}_a(x, a)\rho_a(a, y). \tag{3.269}$$

(v) Using the differential operator D, the derivatives of (3.202) are written as $dP_a(x)/dx = [\mathrm{D}\rho_a B](x)$, $dQ_a(x)/dx = [\mathrm{D}\rho_a A](x)$. We insert commutators so that

$$\frac{dP_a(x)}{dx} = [[\mathrm{D}, \rho_a]B](x) + [\rho_a \mathrm{D}B](x),$$
$$\frac{dQ_a(x)}{dx} = [[\mathrm{D}, \rho_a]A](x) + [\rho_a \mathrm{D}A](x).$$

Applying (3.269) to the above, we obtain

$$\frac{dP_a(x)}{dx} = -\frac{Q_a(x)w(a)}{2t} - \frac{P_a(x)v(a)}{2t} + \mathrm{R}_a(x,a)P_a(a) + \frac{P_a^{(1)}(x)}{2t} - \sqrt{\frac{N}{t}}Q_a(x),$$
$$\frac{dQ_a(x)}{dx} = -\frac{Q_a(x)v(a)}{2t} - \frac{P_a(x)u(a)}{2t} + \mathrm{R}_a(x,a)Q_a(a) - \frac{Q_a^{(1)}(x)}{2t} + \sqrt{\frac{N}{t}}P_a(x), \tag{3.270}$$

where $v(a) = \int_{\mathbb{R}} dy\, P_a(y)\mathbf{1}_{(y\geq a)}A(y) = \int_a^\infty dy\, P_a(y)A(y)$,

$$P_a^{(1)}(x) = \int_{\mathbb{R}} dy\, \rho_a(x,y)yB(y), \quad Q_a^{(1)}(x) = \int_{\mathbb{R}} dy\, \rho_a(x,y)yA(y). \tag{3.271}$$

Note that $v(a)$ is also given by $v(a) = \int_{\mathbb{R}} dy\, Q_a(y)\mathbf{1}_{(y\geq a)}B(y) = \int_a^\infty dy\, Q_a(y)B(y)$. Derive the equalities

$$P_a^{(1)}(x) = xP_a(x) - Q_a(x)w(a) + P_a(x)v(a),$$
$$Q_a^{(1)}(x) = xQ_a(x) - Q_a(x)v(a) + P_a(x)u(a), \tag{3.272}$$

and prove the lemma.

3.24 Prove Lemma 3.13 following the instructions below.
(i) By (3.202),

$$\frac{\partial P_a(x)}{\partial a} = \left[\frac{\partial \rho_a}{\partial a}B\right](x). \tag{3.273}$$

Consider the identity $\rho_a^{-1}\rho_a = (1-\mathrm{K}_a)\rho_a = 1$. Differentiation with respect to a gives $-(\partial \mathrm{K}_a/\partial a)\rho_a + (1-\mathrm{K}_a)(\partial\rho_a/\partial a) = 0$ and thus $\partial\rho_a/\partial a = \rho_a(\partial \mathrm{K}_a/\partial a)\rho_a$. Show that this gives

$$\frac{\partial \rho_a}{\partial a}(x,y) = -\mathrm{R}_a(x,a)\rho_a(a,y). \tag{3.274}$$

Then (3.273) gives

$$\frac{\partial P_a(x)}{\partial a} = -\mathrm{R}_a(x, a) \int_{\mathbb{R}} dz\, \rho_a(a, z) B(z) = -\mathrm{R}_a(x, a) p(a). \tag{3.275}$$

(ii) We should regard $p(a) = P_a(a)$ as $P_a(I(a))$ with $I(a) = a$. Then

$$\begin{aligned} p'(a) &= \frac{dI(a)}{da} \frac{\partial P_a(x)}{\partial x}\bigg|_{x=a} + \frac{\partial P_a(x)}{\partial a}\bigg|_{x=a} \\ &= \frac{\partial P_a(x)}{\partial x}\bigg|_{x=a} + \frac{\partial P_a(x)}{\partial a}\bigg|_{x=a}, \end{aligned}$$

since $I'(a) = 1$. Noting this, prove the lemma.

References

1. Abramowitz, M., Stegun, I.: Handbook of Mathematical Functions. Dover, New York (1965)
2. Adler, M., van Moerbeke, P.: PDFs for the joint distributions of the Dyson, Airy and Sine processes. Ann. Probab. **33**, 1326–1361 (2005)
3. Akemann, G., Baik, J., Di Francesco, P.: The Oxford Handbook of Random Matrix Theory. Oxford University Press, Oxford (2011)
4. Amir, G., Corwin, I., Quastel, J.: Probability distribution of the free energy of the continuum directed random polymer in 1+1 dimensions. Commun. Pure Appl. Math. **64**, 466–537 (2011)
5. Anderson, G.W., Guionnet, A., Zeitouni, O.: An Introduction to Random Matrices. Cambridge University Press, Cambridge (2010)
6. Andraus, S., Katori, M., Miyashita, S.: Interacting particles on the line and Dunkl intertwining operator of type A: application to the freezing regime. J. Phys. A: Math. Theor. **45**, 395201/1-26 (2012)
7. Andraus, S., Katori, M., Miyashita, S.: Two limiting regimes of interacting Bessel processes. J. Phys. A: Math. Theor. **47**, 235201/1-30 (2014)
8. Andrews, G.E., Askey, R., Roy, R.: Special Functions. Cambridge University Press, Cambridge (1999)
9. Arrowsmith, D.K., Mason, P., Essam, J.W.: Vicious walkers, flows and directed percolation. Physica A **177**, 267–272 (1991)
10. Baik, J.: Random vicious walks and random matrices. Commun. Pure Appl. Math. **53**, 1385–1410 (2000)
11. Baik, J., Ben Arous, G., Péché, S.: Phase transition of the largest eigenvalue for nonnull complex sample covariance matrices. Ann. Probab. **33**, 1643–1697 (2005)
12. Baryshnikov, Y.: GUEs and queues. Probab. Theory Relat. Fields **119**, 256–274 (2001)
13. Bertini, L., Giacomin, G.: Stochastic Burgers and KPZ equations from particle systems. Commun. Math. Phys. **183**, 571–607 (1997)
14. Biane, P., Bougerol, P., O'Connell, N.: Littelmann paths and Brownian paths. Duke Math. J. **130**, 127–167 (2005)
15. Biane, P., Bougerol, P., O'Connell, N.: Continuous crystal and Duistermaat-Heckman measure for Coxeter groups. Adv. Math. **221**, 1522–1583 (2009)
16. Bleher, P.M., Kuijlaars, A.B.: Random matrices with external source and multiple orthogonal polynomials. Int. Math. Res. Not. **2004**, 109–129 (2004)
17. Bleher, P.M., Kuijlaars, A.B.: Integral representation for multiple Hermite and multiple Laguerre polynomials. Ann. Inst. Fourier **55**, 2001–2014 (2005)

18. Borodin, A.: Biorthogonal ensembles. Nucl. Phys. **B536**, 704–732 (1999)
19. Borodin, A., Corwin, I.: Macdonald processes. Probab. Theory Relat. Fields **158**, 225–400 (2014)
20. Borodin, A., Rains, E.M.: Eynard-Mehta theorem, Schur process, and their Pfaffian analog. J. Stat. Phys. **121**, 291–317 (2005)
21. Bougerol, P., Jeulin, T.: Paths in Weyl chambers and random matrices. Probab. Theory Relat. Fields **124**, 517–543 (2002)
22. Brézin, E., Hikami, S.: Level spacing of random matrices in an external source. Phys. Rev. E **58**, 7176–7185 (1998)
23. Bru, M.F.: Diffusions of perturbed principal component analysis. J. Multivar. Anal. **29**, 127–136 (1989)
24. Bru, M.F.: Wishart process. J. Theor. Probab. **4**, 725–751 (1991)
25. Calabrese, P., Le Doussal, P., Rosso, A.: Free-energy distribution of the directed polymer at high temperature. Europhys. Lett. **90**, 20002 (2010)
26. Cardy, J., Katori, M.: Families of vicious walkers. J. Phys. A **36**, 609–629 (2003)
27. Cépa, E., Lépingle, D.: Diffusing particles with electrostatic repulsion. Probab. Theory Relat. Fields **107**, 429–449 (1997)
28. Chan, T.: The Wigner semi-circle law and eigenvalues of matrix-valued diffusions. Probab. Theory Relat. Fields **93**, 249–272 (1992)
29. Chhaibi, R.: Littelmann path model for geometric crystals, Whittaker functions on Lie groups and Brownian motion. PhD thesis in Université Paris VI (2013). arXiv:math.PR/1302.0902
30. de Gennes, P.-G.: Soluble model for fibrous structures with steric constraints. J. Chem. Phys. **48**, 2257–2259 (1968)
31. de Haro, S., Tierz, M.: Discrete and oscillatory matrix model in Chern-Simons theory. Nucl. Phys. B **731**, 225–241 (2005)
32. Demni, N.: A guided tour in the world of radial Dunkl processes. In: Graczyk, P., Rösler, M., Yor, M. (eds.) Harmonic and Stochastic Analysis of Dunkl Processes. Herman, Paris (2008)
33. Demni, D.: Radial Dunkl processes: existence, uniqueness and hitting time. C.R. Acad. Sci. Paris, Sér. I Math. **347**, 1125–1128 (2009)
34. Doob, J.L.: Classical Potential Theory and its Probabilistic Counterpart. Springer, Berlin (1984)
35. Dotsenko, V.: Replica Bethe ansatz derivation of the Tracy-Widom distribution of the free energy fluctuations in one-dimensional directed polymers. J. Stat. Mech. P07010 (2010)
36. Dotsenko, V., Klumov, B.: Bethe ansatz solution for one-dimensional directed polymers in random media. J. Stat. Mech. P03022 (2010)
37. Doumerc, Y., Moriarty, J.: Exit problems associated with affine reflection groups. Probab. Theory Relat. FIelds **145**, 351–383 (2009)
38. Dyson, F.J.: A Brownian-motion model for the eigenvalues of a random matrix. J. Math. Phys. **3**, 1191–1198 (1962)
39. Esaki, S.: Nocolliding system of continuous-time random walks. Pacific J. Math. Indus. **6**, 11/1-10 (2014)
40. Eynard, B., Mehta, M.L.: Matrices coupled in a chain: I Eigenvalue correlations. J. Phys. A **31**, 4449–4456 (1998)
41. Feierl, T.: The height of watermelon with wall. J. Phys. A **45**, 095003/1-26 (2012)
42. Ferrari, P.L., Spohn, H.: Constrained Brownian motion: fluctuations away from circular and parabolic barriers. Ann. Probab. **33**, 1302–1325 (2005)
43. Fisher, M.E.: Walks, walls, wetting, and melting. J. Stat. Phys. **34**, 667–729 (1984)
44. Forrester, P.J.: The spectrum edge of random matrix ensemble. Nucl. Phys. B **402** [FS], 709–728 (1993)
45. Forrester, P.J.: Log-gases and random matrices. London Mathematical Society Monographs. Princeton University Press, Princeton (2010)
46. Forrester, P.J., Nagao, T., Honner, G.: Correlations for the orthogonal-unitary and symplectic-unitary transitions at the hard and soft edges. Nucl. Phys. B **553** [PM], 601–643 (1999)

47. Fukushima, M., Oshima, Y., Takeda, M.: Dirichlet Forms and Symmetric Markov Processes. Walter de Gruyter, Berlin (1994)
48. Fulton, W.: Young Tableaux with Applications to Representation Theory and Geometry. Cambridge University Press, Cambridge (1997)
49. Funaki, T., Quastel, J.: KPZ equation, its renormalization and invariant measures. Stoch. PDE: Anal. Comp. **3**, 159–220 (2015)
50. Gessel, I., Viennot, G.: Binomial determinants, paths, and hook length formulae. Adv. Math. **58**, 300–321 (1985)
51. Grabiner, D.J.: Brownian motion in a Weyl chamber, non-colliding particles, and random matrices. Ann. Inst. Henri Poincaré, Probab. Stat. **35**, 177–204 (1999)
52. Graczyk, P., Małecki, J.: Multidimensional Yamada-Watanabe theorem and its applications to particle systems. J. Math. Phys. **54**, 021503/1-15 (2013)
53. Graczyk, P., Małecki, J.: Strong solutions of non-colliding particle systems. Electron. J. Probab. **19**(119), 1–21 (2014)
54. Gravner, J., Tracy, C.A., Widom, H.: Limit theorems for height fluctuations in a class of discrete space and time growth models. J. Stat. Phys. **102**, 1085–1132 (2001)
55. Hairer, M.: Solving the KPZ equation. Ann. Math. **178**, 559–664 (2013)
56. Hastings, S.P., McLeod, J.B.: A boundary value problem associated with the second Painlevé transcendent and the Korteweg-de Vries equation. Arch. Rat. Mech. Anal. **73**, 31–51 (1980)
57. Hough, J.B., Krishnapur, M., Peres, Y., Virág, B.: Zeros of Gaussian Analytic Functions and Determinantal Point Processes. University Lecture Series. Amer. Math. Soc. **51**, Providence (2009)
58. Ikeda, N., Watanabe, S.: Stochastic Differential Equations and Diffusion Processes, 2nd edn. North-Holland/Kodansha, Tokyo (1989)
59. Imamura, T., Sasamoto, T.: Fluctuations of the one-dimensional polynuclear growth model with external sources. Nucl. Phys. B **699**, 503–544 (2004)
60. Ismail, M.E.H.: Classical and Quantum Orthogonal Polynomials in One Variable. Cambridge University Press, Cambridge (2005)
61. Izumi, M., Katori, M.: Extreme value distributions of noncolliding diffusion processes. RIMS Kôkyûroku Bessatsu **B27**, 45–65 (2011)
62. Jimbo, M., Miwa, T., Môri, Y., Sato, M.: Density matrix of an impenetrable Bose gas and the fifth Painlevé transcendent. Physica D **1**, 80–158 (1980)
63. Johansson, K.: Universality of the local spacing distribution in certain ensembles of Hermitian Wigner matrices. Commun. Math. Phys. **215**, 683–705 (2001)
64. Johansson, K.: Non-intersecting paths, random tilings and random matrices. Probab. Theory Relat. Fields **123**, 225–280 (2002)
65. Johansson, K.: Discrete polynuclear growth and determinantal processes. Commun. Math. Phys. **242**, 277–329 (2003)
66. Kardar, M., Parisi, G., Zhang, Y.C.: Dynamic scaling of growing interfaces. Phys. Rev. Lett. **56**, 889–892 (1986)
67. Karlin, S., McGregor, J.: Coincidence probabilities. Pacific J. Math. **9**, 1141–1164 (1959)
68. Katori, M.: O'Connell's process as a vicious Brownian motion. Phys. Rev. E **84**, 061144/1-11 (2011)
69. Katori, M.: Determinantal process starting from an orthogonal symmetry is a Pfaffian process. J. Stat. Phys. **146**, 249–263 (2012)
70. Katori, M.: Survival probability of mutually killing Brownian motion and the O'Connell process. J. Stat. Phys. **147**, 206–223 (2012)
71. Katori, M.: Reciprocal time relation of noncolliding Brownian motion with drift. J. Stat. Phys. **148**, 38–52 (2012)
72. Katori, M.: System of complex Brownian motions associated with the O'Connell process. J. Stat. Phys. **149**, 411–431 (2012)
73. Katori, M.: Determinantal martingales and noncolliding diffusion processes. Stoch. Proc. Appl. **124**, 3724–3768 (2014)

74. Katori, M.: Determinantal martingales and correlations of noncolliding random walks. J. Stat. Phys. **159**, 21–42 (2015)
75. Katori, M.: Elliptic determinantal process of type A. Probab. Theory Relat. Fields **162**, 637–677 (2015)
76. Katori, M.: Characteristic polynomials of random matrices and noncolliding diffusion processes. RIMS Kôkyûroku **1970**, 22–44 (2015)
77. Katori, M., Nagao, T., Tanemura, H.: Infinite systems of non-colliding Brownian particles. Adv. Stud. Pure Math. 39. Stochastic Analysis on Large Scale Interacting Systems, pp. 283–306. Mathematical Society of Japan, Tokyo (2004)
78. Katori, M., Tanemura, H.: Scaling limit of vicious walks and two-matrix model. Phys. Rev. E **66**, 011105/1-12 (2002)
79. Katori, M., Tanemura, H.: Functional central limit theorems for vicious walkers. Stoch. Stoch. Rep. **75**, 369–390 (2003)
80. Katori, M., Tanemura, H.: Symmetry of matrix-valued stochastic processes and noncolliding diffusion particle systems. J. Math. Phys. **45**, 3058–3085 (2004)
81. Katori, M., Tanemura, H.: Infinite systems of noncolliding generalized meanders and Riemann-Liouville differintegrals. Probab. Theory Relat. Fields **138**, 113–156 (2007)
82. Katori, M., Tanemura, H.: Noncolliding Brownian motion and determinantal processes. J. Stat. Phys. **129**, 1233–1277 (2007)
83. Katori, M., Tanemura, H.: Zeros of Airy function and relaxation process. J. Stat. Phys. **136**, 1177–1204 (2009)
84. Katori, M., Tanemura, H.: Non-equilibrium dynamics of Dyson's model with an infinite number of particles. Commun. Math. Phys. **293**, 469–497 (2010)
85. Katori, M., Tanemura, H.: Noncolliding squared Bessel processes. J. Stat. Phys. **142**, 592–615 (2011)
86. Katori, M., Tanemura, H.: Noncolliding processes, matrix-valued processes and determinantal processes. Sugaku Expositions (AMS) **24**, 263–289 (2011)
87. Katori, M., Tanemura, H.: Complex Brownian motion representation of the Dyson model. Electron. Commun. Probab. **18**(4), 1–16 (2013)
88. Kobayashi, N., Izumi, M., Katori, M.: Maximum distributions of bridges of noncolliding Brownian paths. Phys. Rev. E **78**, 051102/1-15 (2008)
89. König, W., O'Connell, N.: Eigenvalues of the Laguerre process as non-colliding squared Bessel process. Elec. Commun. Probab. **6**, 107–114 (2001)
90. Krattenthaler, C.: Advanced determinant calculus. Séminaire Lotharingien Combin. 42 (The Andrews Festschrift), B42q (1999)
91. Krattenthaler, C., Guttmann, A.J., Viennot, X.G.: Vicious walkers, friendly walkers and Young tableaux: II. With a wall. J. Phys. A : Math. Gen. **33**, 8835–8866 (2000)
92. Levin, B.Y.: Lectures on Entire Functions. Translations of Mathematical Monographs, vol. 150. Amer. Math. Soc., Providence (1996)
93. Lindström, B.: On the vector representations of induced matroids. Bull. London Math. Soc. **5**, 85–90 (1973)
94. Macdonald, I.G.: Some conjectures for root systems. SIAM J. Math. Anal. **13**, 988–1007 (1982)
95. Macdonald, I.G.: Symmetric Functions and Hall Polynomials, 2nd edn. Oxford University Press, Oxford (1995)
96. Mariño, M.: Chern-Simons theory, matrix integrals, and perturbative three-manifold invariants. Commun. Math. Phys. **253**, 25–49 (2005)
97. Mariño, M.: Chern-Simons Theory, Matrix Models and Topological Strings. Oxford University Press, Oxford (2005)
98. Matsumoto, S., Shirai, T.: Correlation functions for zeros of a Gaussian power series and Pfaffians. Electron. J. Probab. **18**(49), 1–18 (2013)
99. Matsumoto, H., Yor, M.: An analogue of Pitman's $2M - X$ theorem for exponential Wiener functionals, Part I: A time-inversion approach. Nagoya Math. J. **159**, 125–166 (2000)

100. Matsumoto, H., Yor, M.: Exponential functionals of Brownian motion I: Probability laws at fixed time. Probab. Surv. **2**, 312–347 (2005)
101. Mehta, M.L.: Random Matrices, 3rd edn. Elsevier, Amsterdam (2004)
102. Muttalib, K.A.: Random matrix models with additional interactions. J. Phys. A **28**, L159–L164 (1995)
103. Nagao, T.: Dynamical correlations for vicious random walk with a wall. Nucl. Phys. B **658** [FS], 373–396 (2003)
104. Nagao, T.: Pfaffian expressions for random matrix correlation functions. J. Stat. Phys. **129**, 1137–1158 (2007)
105. Nagao, T., Forrester, P.: Multilevel dynamical correlation functions for Dyson's Brownian motion model of random matrices. Phys. Lett. A **247**, 42–46 (1998)
106. Nagao, T., Katori, M., Tanemura, H.: Dynamical correlations among vicious random walkers. Phys. Lett. A **307**, 29–35 (2003)
107. O'Connell, N.: Directed polymers and the quantum Toda lattice. Ann. Probab. **40**, 437–458 (2012)
108. O'Connell, N., Yor, M.: A representation for non-colliding random walks. Electron. Commun. Probab. **7**, 1–12 (2002)
109. Osada, H.: Dirichlet form approach to infinite-dimensional Wiener processes with singular interactions. Commun. Math. Phys. **176**, 117–131 (1996)
110. Osada, H.: Infinite-dimensional stochastic differential equations related to random matrices. Probab. Theory Relat. Fields **153**, 471–509 (2012)
111. Osada, H.: Interacting Brownian motions in infinite dimensions with logarithmic interaction potentials. Ann. Probab. **41**, 1–49 (2013)
112. Osada, H.: Interacting Brownian motions in infinite dimensions with logarithmic interaction potentials II: Airy random point field. Stoch. Proc. Appl. **123**, 813–838 (2013)
113. Pitman, J.W.: One-dimensional Brownian motion and the three-dimensional Bessel process. Adv. Appl. Prob. **7**, 511–526 (1975)
114. Prähofer, M., Spohn, H.: Scale invariance of the PNG droplet and the Airy process. J. Stat. Phys. **108**, 1071–1106 (2002)
115. Prähofer, M., Spohn, H.: Exact scaling functions for one-dimensional stationary KPZ growth. J. Stat. Phys. **115**, 255–279 (2004). http://www-m5.ma.tum.de/KPZ
116. Revuz, D., Yor, M.: Continuous Martingales and Brownian Motion, 3rd edn. Springer, Berlin (1999)
117. Rogers, L.C.G., Shi, Z.: Interacting Brownian particles and the Wigner law. Probab. Theory Relat. Fields **95**, 555–570 (1993)
118. Sasamoto, T.: Spatial correlations of the 1D KPZ surface on a flat substrate. J. Phys. A **38**, L549–L556 (2005)
119. Sasamoto, T., Spohn, H.: The crossover regime for the weakly asymmetric simple exclusion process. J. Stat. Phys. **128**, 799–846 (2010)
120. Sasamoto, T., Spohn, H.: Exact height distributions for the KPZ equation with narrow wedge initial condition. Nucl. Phys. B **834**, 523–542 (2010)
121. Schehr, G., Majumdar, S.N., Comtet, A., Randon-Furling, J.: Exact distribution of the maximum height of p vicious walkers. Phys. Rev. Lett. **101**, 150601/1-4 (2008)
122. Selberg, A.: Bemerkninger om et multiplet integral. Nor. Matematisk Tidsskr. **26**, 71–78 (1944)
123. Shirai, T., Takahashi, Y.: Random point fields associated with certain Fredholm determinants I: fermion, Poisson and boson point processes. J. Funct. Anal. **205**, 414–463 (2003)
124. Shirai, T., Takahashi, Y.: Random point fields associated with certain Fredholm determinants II: fermion shifts and their ergodic and Gibbs properties. Ann. Probab. **31**, 1533–1564 (2003)
125. Simon, B.: Trace ideals and their applications, 2nd edn. Amer. Math. Soc. Providence (2005)
126. Soshnikov, A.: Determinantal random point fields. Russ. Math. Surv. **55**, 923–975 (2000)
127. Spohn, H.: Interacting Brownian particles: a study of Dyson's model. In: Papanicolaou, G. (ed.) Hydrodynamic Behavior and Interacting Particle Systems. IMA Volumes in Mathematics and its Applications, vol. 9, pp. 151–179. Springer, Berlin (1987)

128. Stanley, R.P.: Enumerative Combinatorics, vol. 2. Cambridge University Press, Cambridge (1999)
129. Szegö, G.: Orthogonal Polynomials, 4th edn. Amer. Math. Soc, Providence (1975)
130. Takahashi, Y., Katori, M.: Noncolliding Brownian motion with drift and time-dependent Stieltjes-Wigert determinantal point process. J. Math. Phys. **53**, 103305/1-23 (2012)
131. Takahashi, Y., Katori, M.: Oscillatory matrix model in Chern-Simons theory and Jacobi-theta determinantal point process. J. Math. Phys. **55**, 093302/1-24 (2014)
132. Tanemura, H.: A system of infinitely many mutually reflecting Brownian balls on $\mathbb{R}^d$. Probab. Theory Relat. Fields **104**, 399–426 (1996)
133. Tao, T.: Topics in Random Matrix Theory. Amer. Math. Soc, Providence (2012)
134. Tierz, M.: Soft matrix models and Chern-Simons partition functions. Mod. Phys. Lett. A **19**, 1365–1378 (2004)
135. Titchmarsh, E.C.: Eigenfunction Expansions Associated with Second-Order Differential Equations. Part I, 2nd edition, Clarendon Press, Oxford (1962)
136. Tracy, C.A., Widom, H.: Level-spacing distributions and the Airy kernel. Commun. Math. Phys. **159**, 151–174 (1994)
137. Tracy, C.A., Widom, H.: Fredholm determinants, differential equations and matrix models. Commun. Math. Phys. **163**, 33–72 (1994)
138. Tracy, C.A., Widom, H.: The distribution of the largest eigenvalue in the Gaussian ensembles: $\beta = 1, 2, 4$. In: van Diejen, J.F., Vinet L. (eds.) Calogero-Moser-Sutherland Models. CRM Series in Mathematical Physics, vol. 4, pp. 461–472. Springer, New York (2000)
139. Tracy, C.A., Widom, H.: A system of differential equations for the Airy process. Electron Commun. Probab. **8**, 93–98 (2003)
140. Tracy, C.A., Widom, H.: Differential equations for Dyson processes. Commun. Math. Phys. **252**, 7–41 (2004)
141. Tracy, C.A., Widom, H.: Asymptotics in ASEP with step initial condition. Commun. Math. Phys. **290**, 129–154 (2009)
142. Vallée, O., Soares, M.: Airy Functions and Applications to Physics. Imperial College Press, London (2004)
143. Warren, J.: Dyson's Brownian motions, intertwining and interlacing. Electron. J. Probab. **12**(19), 573–590 (2007)
144. Yoo, H.J.: Dirichlet forms and diffusion processes for fermion random point fields. J. Funct. Anal. **219**, 143–160 (2005)
145. Zinn-Justin, P.: Universality of correlation functions of Hermitian random matrices in an external field. Commun. Math. Phys. **194**, 631–650 (1998)

Index

M. Katori, *Bessel Processes, Schramm–Loewner Evolution, and the Dyson Model*, SpringerBriefs in Mathematical Physics 11, DOI 10.1007/978-981-10-0275-5

Printed in the United States
By Bookmasters